AF632559

Climate Smart Practices

NIPA® GENX ELECTRONIC RESOURCES & SOLUTIONS P. LTD.
New Delhi-110 034

About the Editor

Prof. Kalyan Ghadei Ph.D(BHU), B.Mus (ABGM-M),C.C(PI) has a good academic career throughout . Prof. Ghadei has successfully recorded activities at All India Radio, Cuttack, CIFA(ICAR), Bhubaneswar and in OUAT Odisha for transfer of Agricultural, aquaculture and pearl culture technology in Odisha and India. He is serving as a Professor & Head in the Department of Extension Education in Banaras Hindu University, BHU, Varanasi. He has published more than 60 research papers , authored 09 books and about 25 book chapters. He is a member of many national and international professional societies. Prof. Ghadei has received many awards of national and international repute. The most significant award to his credit is Raman fellowship as a visiting scholar to USA and Post Doctoral fellow at Virginia Tech , USA to study climate change issues and its mitigation strategies.

Climate Smart Practices At Grassroot Levels

Editor
Kalyan Ghadei

Co-Editor
Nikhil J.
Jayashree M.P.
Suman Roy
Waris Ali

NIPA® GENX ELECTRONIC RESOURCES & SOLUTIONS P. LTD.
New Delhi-110 034

NIPA. GENX ELECTRONIC
RESOURCES & SOLUTIONS P. LTD.

101,103, Vikas Surya Plaza, CU Block
L.S.C.Market, Pitam Pura, New Delhi-110 034
Ph : +91 11 4386 0225, 9717133558, 9540816132
E-mail: newindiapublishingagency@gmail.com
Website: www. niparesources.com

© 2024, Publisher

Print ISBN: 978-93-58879-45-2

ebook ISBN: 978-93-58877-21-2

All rights reserved. No part of this publication may be reproduced, stored in a retrieval system or transmitted in any form or by any means, including electronic, mechanical, photocopying recording or otherwise without the prior written permission of the publisher or the copyright holder.

This book contains information obtained from authentic and highly reliable sources. Reasonable efforts have been made to publish reliable data and information, but the author/s, editor/s and publisher cannot assume responsibility for the validity, accuracy or completeness of all materials or information published herein or the consequences of their use. The work is published with the understanding that the publisher and author/s are not attempting to render any professional services. The author/s, editor/s and publisher have attempted to trace and acknowledge the copyright holders of all material reproduced in this publication and apologize to copyright holders if permission and/or acknowledgements to publish in this form have not been taken. If any copyrighted material has not been acknowledged, please write to us and let us know so that we may rectify the error, in subsequent reprints.

Trademark Notice: NIPA®, the NIPA® logos and their presentations (the way they are written/presented) in this book are the trademarks of the publisher and hence may not be used without written permission, if copied or used without authorization, the infringer will be prosecuted as per law.

NIPA® also publishes books in a variety of electronic formats. Some content that appears in print may not be available in electronic books, and vice versa

Composed and Designed by NIPA®.

Preface

Agriculture being sensitive to changes in the climate, the climate change can have adverse impact on food production and pose prominent threat to food security to the countries in particular and to the world population in general. Climate change refers to any change in climate overtime, whether due to natural variability or as a result of human activity (inter-governmental panel on climate change, IPCC, 2001). It can also be seen as change in climate which is attributed directly or indirectly to human activities that alter the composition of the global atmosphere and which are in addition to natural variability observed over comparable time periods (IPCC, 2007). Climate change has become a global issue in recent times manifesting in variations of different climate parameters including cloud cover, precipitation, temperature ranges, sea levels and significant effects through the production and release of GHGs.

Agricultural extension according to Leeuwis (2006) is a series of embedded communicative interventions that are meant, among other things, to develop and/or induce innovations which supposedly help to resolve (usually multi-actor) problematic situations. It has been observed that agricultural extension is involved in public information and education programmes that could assist farmers in mitigating the effects of climate change (MOE FRN, 2003). According to them, such involvements include awareness creation and knowledge brokerage on the issues of climate change; building resilience capacities among vulnerable individuals, communities and regions; encouragement of wide participation of all stakeholders in addressing climate change issues; and developing appropriate frameworks for coping/adapting to climate change effects/impacts.

We would like to extend our heartfelt gratitude to the esteemed contributors who have shared their insights and expertise, making this publication possible. Their unwavering commitment to pushing the boundaries of knowledge has enriched this book immensely.

We are also deeply appreciative of the countless volunteers and supporters, both directly and indirectly involved in this endeavor. Your assistance, whether through guidance, encouragement, or practical support, has been instrumental in bringing this manuscript to fruition.

A special thanks is owed to the associate editors and research scholars of the Department of Agricultural Extension Education at Banaras Hindu University. Your relentless dedication to the cause of agricultural extension and your unwavering support in the development of this book are sincerely appreciated.

As we embark on this journey through the pages of Climate Change and Agricultural Extension, we encourage readers to explore the depths of knowledge contained herein. It is our hope that this book will not only inform but also inspire new generations of entrepreneurs and extension professionals to shape a more vibrant and prosperous future for agriculture and rural development.

Kalyan Ghadei

Contents

1

Role of Extension Functionaries in Adoption of Natural Farming

Minam Gamoh, Brij Bala, Komal Sharma and Rajesh Kumar Thakur

Chaudhary Sarwan Kumar Himachal Pradesh Krishi Vishvavidyalaya Palampur, Himachal Pradesh

Abstract

India attained self sufficiency in foodgrain production and emerged as major exporter of food grains through the use of conventional agricultural production technologies. But the excessive and indiscriminate use of chemicals in crop production not only adversely affected the soil health but also increased the cost of production. In recent years the need has been felt to develop alternate technologies and practices which are sustainable and climate resilient. In this context, Sh. Subhash Palekar was working on these lines since 1990s and came up with module of farming, popularly known as Subash Palekar Natural Farming (SPNF). The practices are based on inputs which are of farm origin and the production is sustainable in nature. The SPNF is being promoted in Himachal Pradesh since 2018 under "PrakritikKheti Khushhal Kissan" scheme and nearly 54000 farmers has partially adopted SPNF by 2019-20. In this article an effort has been made to examine the role of extension functionaries in the adoption of SPNF in the state. Garrett ranking technique was used to find out the factors that makes the training on natural farming effective. Primary data was collected from 80 SPNF farmers through two stage random sampling technique in the agricultural year 2021-22. The study revealed that majority of respondents (83.75%) were trained, out of which 64.18, 23.88, 7.46 and 4.48 per cent were motivated by Agricultural Technology Management Agency (ATMA), themselves, neighbors and relatives, respectively. Introduction of effective pesticide to natural farming was the most important factor that will make the training effective followed by training on package of practices, inclusion of soil and water testing facility, supply of quality seeds, more field demonstration and hands on approach, timely visit by official and more research on fruit crops.

Keywords: Natural Farming, Sustainability, Extension functionaries, Training, Adoption, Garrett ranking

Introduction

India was a self-sufficient nation in its food production before independence but after the occurrence of various famines, the government in late 1960s started focusing more on the crop production *i.e.,*green revolution (NAAS 2019). Use of hybrids, fertilizers and plant protection chemicals increased and major investments were made on irrigation structures which led to a massive jump in the production of crops. But with time, the negative externalities caused by green revolution were vividly noticeable. The indiscriminate use of chemicals has debilitated the environment by causing soil depletion, pollution, deforestation, climate change *etc.* thus affecting the health of human kind. The lofty prices of inputs led the farmers to resort to debt thus falling into debt trap and suicides. In order to address this problem an alternative to the conventional farming is required which is more sustainable and climate resilient. In the mid-1990s, natural farming *viz.*, Zero Budget Natural Farming (ZBNF) was demonstrated by Sh. Subhash Palekar also known Subash Palekar Natural Farming (SPNF). It is an agroecology based diversified farming system which involves no financial outlay to purchase essential from outside and makes farming profitable. Mishra (2018) studied on Zero Budget Natural Farming (ZBNF) and found that crops grown under ZBNF evinced higher yield than the crops grown under non-ZBNF. The study found that the farmers who have transitioned to natural farming experienced a decrease in the costs of inputs and higher net income as compared to non-ZBNF. Similarly, Byravan (2018) stated in a study of Andhra Pradesh that the yield of food crops and different cash crops in ZBNF were found to be significantly higher in comparison to chemical farming. Tripathi et al. (2018) conducted a study on Zero Budget Natural Farming (ZBNF) for the Sustainable Development Goals (SDGs) in the state of Andhra Pradesh. They stated that ZBNF could assist Andhra Pradesh and India in making a momentous progress towards achieving almost one-fourth of the 169 SDGs targets. They also found that paddy farmers achieved a 51 per cent increase in their net income in comparison to their non-ZBNF counterparts. Commercial crops namely groundnut and cotton showed an increase of net income by 135 per cent and 87 per cent respectively. Galab et al. (2019) studied the impact of Zero Budget Natural Farming (ZBNF) for *Kharif* season in the state of Andhra Pradesh. They found that the cost of cultivation of maize under ZBNF was Rs 13036 per acre which was less than the traditional method of cultivation *i.e.,* 14135 per acre. The production cost under ZBNF was Rs 624 per quintal which was lesser than the non-ZBNF of Rs 824 per quintal. The net returns were Rs 18362 per acre and Rs 8684 per acre for ZBNF and non-ZBNF respectively.

Bishnoi and Bhati (2017) gave an overview of the Zero Budget Natural Farming (ZBNF) stating that the conventional system of farming is a slow poison to the people in terms of human health and also the soil. It turns the soil into a barren land and farmers eventually get trapped in the cycle of debt leading to suicides. Sulok *et al.* (2018) conducted a study in Malaysia to know the effect of introducing natural farming in black pepper cultivation. The study found that the introduction of natural farming in black pepper cultivation has led to the increase in the number of effective micro-organisms in the soil. The growth and yield of the crop could get increased by 25 per cent. This helped in reduction of expenses on chemical inputs along with reduction of environmental damage caused by fungicides, insecticides, pesticides and chemical fertilizers. Similarly, Keerthi et al. (2018) conducted a study on Zero Budget Natural Farming (ZBNF) and found that with the limited use of external inputs and inclusion of supplements *viz.,*Jivamrit, Ghanjeevamrit under natural farming the soil microorganisms and soil productivity improved. The ZBNF also advocates the conservation of micro and macro-flora wit inclusion of multi-cropping as well. Palekar (2005) in his research on Zero Budget Natural Farming (ZBNF) showed that the minimal use of external inputs and use of natural product namely Jivamrit as soil supplements by natural farmers rejuvenate the soil productivity, increase the nutrient availability by increasing the number of soil micro-organisms. The ZBNF advocates multi-cropping and conservation of micro and macro-flora. The cost of production has been reduced by this method.

Himachal Pradesh is an agrarian state in which about 69 per cent of the population gets direct employment from agriculture and its allied sectors. In 2018, the Government of Himachal Pradesh launched the"*PrakritikKheti Khushhal Kissan*" to promote Natural farming in the state to reduce cost of cultivation and enhanced farm income (PKKKY 2022). The agriculture department, Krishi Vigyan Kendras (KVKs), agriculture extension agents played a major role in its implementation. So far, a total of 53,951 farmers have adopted SPNF Model (partially) as against the target of 50,000 in all the 12 districts of the state in 2019-20. Subsidies were provided for lining of 1053 cowsheds of farmers practicing SPNF. About 60,000 copies of pictorials SPNF practices were printed and distributed among farmers. Therefore, to understand the role of extension agents in implementing natural farming, the study was carried out in the state of Himachal Pradesh.

Methodology

Study Area

The Study was conducted in Himachal Pradesh which is located at latitude of $30^0$22'40"N to $33^0$12'40"N and longitude of $75^0$45'55"E to $79^0$04'20"E (Glance 2022). The geographical area of the state is 55673 sq km. the state comprises of 12 districts with a total population of 68.64 lakhs. The population density in the state is 123 persons per sq. km.

Data and Data Sources

To attain the objective, secondary data were collected from government website in the year 2020-2021. Primary data were collected purposively from two agro-climatic zones *i.e.,* Zone-I and Zone-II of the state. Two stage random sampling technique was used for the selection of the blocks and farmers. A total of eight blocks were covered comprising four blocks from each zone and a total of 80 farmers were covered who were practicing natural farming.

Analytical Tools

Techniques like averages and percentages were used to draw inferences. To understand the factors that make the training effective, scoring technique by Garrett was used for converting ranks into scores when the preference of a particular factor differs from respondent to respondent. At first, effort was made to identify the factors that will help to make the training effective. The farmers were asked to rank various factors according to their importance which was then converted into a score value and rank using the following formula:

$$\text{Per cent position} = \frac{100 \times (R_{ij} - 0.5)}{N_j}$$

Where,

Rij = Rank given for the ith variable by jth respondents

Nj = Number of variable ranked by jth respondents

The chart provided by Garrett was then used to convert the per cent position into Garrett values/scores(Garrett and Woodworth 1969) . Then for each factor, the scores of each individual were added and then total value of scores and average values of score was calculated. The factors having highest average score was considered to be the most important factor.

Results

Role of extension functionaries in adoption of Natural Farming

Table 1: Famers Adopting Natural Farming

Particulars	Number	Average duration (days)
1. Farmers received training	67 (83.75%)	5.68
2. Who motivated to adopt Natural Farming		
	Percent	
a. Self	23.88	
b. Neighbor	7.46	
c. Relatives	4.48	
a. Agricultural Technology Management Agency (ATMA)	64.18	

Table 1 revealed that majority of the farmers (83.75%) received SPNF training for an average duration of about 5.68 days. The trainings were organized by state agriculture department and agricultural university. Some of the farmers attended workshops in other states organized by Subash Palekar himself. The study showed that Agricultural Technology Management Agency (ATMA) was the major source (64.18%) that motivated the farmers to adopt Natural Farming followed by Self (23.88%), Neighbor(7.46%) and relatives (4.48%).

Application of Garrett Ranking

Table 2 showed the factors that will make the training effective organized by ATMA. The respondents stated that inclusion of these factors would make the training more effective. Effective training practices will also act as a stimulator for conventional farmers to shift to natural farming. About 37 respondents ranked the need of effective bio-pesticides as first followed by 16 respondents who ranked it as second. Only 2 respondents gave it the least priority thus ranked it in sixth position. More research on fruit crops was given the least importance by majority (49) of the respondents. It was due to the less number of the sample respondents owning orchard.

Table 2: Rank estimation of factors that will make training effective

Factors	1st	2nd	3rd	4th	5th	6th	7th
More research on fruit crops	4	9	1	4	0	0	49
Need of effective bio-pesticides	37	16	1	2	9	2	0
Training on package of practices	9	26	15	6	5	6	0
Supply of quality seeds	7	6	3	16	8	19	8
Inclusion of soil and water testing facility	0	4	19	14	17	13	0
More visit by the official	1	3	12	13	9	20	9
More field demonstration and practical learning	9	3	16	12	19	7	1

Table 3 showed the percent value and Garrett value position. The percent value position is used to rank the exact factor for a specific phenomenon. The chart provided by Garrett was used to convert the per cent position into Garrett value.

Table 3: Percent position and Garrett value

Sl.No.	100(Rij-0.5)/Nj	Percent position	Garette value
1	100(1-0.5)/7	7.14	78
2	100(2-0.5)/7	21.43	66
3	100(3-0.5)/7	35.71	57
4	100(4-0.5)/7	50.00	50
5	100(5-0.5)/7	64.29	43
6	100(6-0.5)/7	78.57	34
7	100(7-0.5)/7	92.86	22

Table 4: Calculation of Garrett value and ranking

	Ist	2nd	3rd	4th	5th	6th	7th	Total	Average score	Rank
More research on fruit crops	312	594	57	200	0	0	1078	2241	33.45	7
Need of effective bio-pesticides	2886	1056	57	100	387	68	0	4554	67.97	1
Training on package of practices	702	1716	855	300	215	204	0	3992	59.58	2
Supply of quality seeds	546	396	171	800	344	646	176	3079	45.96	4
Inclusion of soil and water testing facility	0	264	1083	700	731	442	0	3220	48.06	3
More visit by the official	78	198	684	650	387	680	198	2875	42.91	6
More field demonstration and practical learning	702	198	912	0	817	238	22	2889	43.12	5

Table 4 showed that need of effective bio pesticides (67.97) was the most important factor with the maximum garrett score of 4554 that will help in the improvement of training. The respondents were not satisfied with the effectiveness of the bio-pesticides. Therefore, introduction of effective bio -pesticides to the farmers while carrying out various training programmes will make the training more effective. Training on package of practices (59.58) was the second important factor with an average score of 3992 that will make the training effective as some farmers were not well acquainted with the package of practices *viz.,* sowing of seeds in intecropping, mulching, spraying of natural inputs*etc.*Inclusion of soil and water testing facility was ranked third with Garrett score of 3220 and average score of 48.06 as most of the respondents expressed disappointment that there is no provision of soil and water testing of their field. Some farmers feared the negative impact on their crops due to the entrance of irrigation water from neighboring field that were chemically treated. As the fields in villages are located very close to each other there is a high chance of entrance of the residues from one field to another. Many farmers used their own seeds for sowing and some received from the department but supply of quality seeds was ranked the fourth (45.96) among the seven factors. More field demonstration was ranked fifth with an average score of 43.12. The farmers expressed that more demonstration in field and practical learning would be of great help for them as they believed that the trainings lack practical lessons. A timely visit by official was ranked the sixth as the visits will help the farmers to assess their progress and improve in the practice. An average score of 42.91 was given with garrett score of 2875. The farmers who owned orchards revealed that in natural farming less work has been done for the fruit crops as more focus has been made in vegetable and cereal crops. This factor was ranked last with the least average score of 33.45 and a Garrett score of 2241.

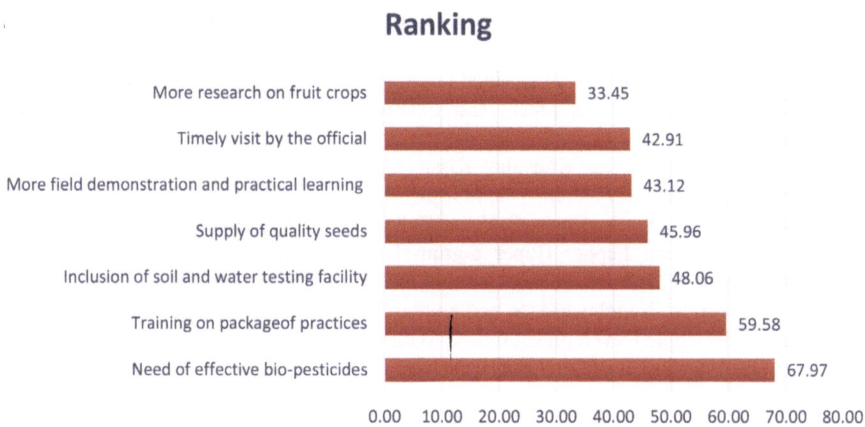

Fgure 1: Ranking offactors

Conclusion

From the study it was observed that ATMA plays a major role in the adoption of natural farming by the farmers in the state. Need of effective bio-pesticide was the major factor that will make the training more effective as stated by the respondents. This would also encourage more farmers to shift to natural farming. Therefore, policy makers should focus more on research for preparation of effective bio-pesticides. Timely visits and assessment of the technology should be done in the farm itself at least for the initial stage of the programme implementation until farmers are confident enough to carry out the practices themsleves.

References

Bishnoi, R. and Bhati, A. (2017). An Overview : Zero Budget Natural Farming. Trends in Biosciences 10(46).

Byravan, S. (2018). The seeds of sustainability. p. 57.https://www.thehindu.com/opinion/op-ed/the- seeds-of-sustainability/article24213217.ece.

Galab, S., Reddy, P.P., Raju, S.R., Ravi, C. and Rajani, A. (2019). Impact Assessment of Zero Budget Natural Farming in Andhra Pradesh- Kharif 2018-19. A comprehensive approach using crop cutting and experiments. Centre for Economic and Social Studies. Nizamiah Observatory Campus, Begumpet, Telangana, India. pp. 45.

Garrett, E. H and Woodworth, R. S. (1969). Statistics in psychology and education. Vakils, Feffer and Simons Pvt. Ltd., Bombay, Pp. 329.

Glance 2022. At a glance. https://himachal.nic.in/en-IN/at-a-glance.html

Keerthi, P., Sharma, S.K. and Chaudhary, K. (2018). Zero Budget Natural Farming: An introduction to Zero Budget Natural Farming In: Research Trends in Agriculture Sciences. 13: 113-123.

Mishra, S. (2018). Zero Budget Natural Farming. Are this and similar practices the answers, Nabakrushna Choudhary Centre For Development Studies, Bhubaneswar, Odisha. 21-23.

NAAS 2019. Zero Budget Natural Farming - A Myth or Reality? Policy Paper No. 90, National Academy of Agricultural Sciences, New Delhi: 20.

Palekar, S. (2005). The philosophy of spiritual farming I. 2nd ed. Amravati: Zero Budget Natural Farming Research, Development and Extension Movement, Amravati, Maharastra, India.

PKKKY. (2022). PrakritikKheti Khushhal Kisan Yojana. State Project Implementing Unit. https://spnfhp.nic.in/SPNF/en-IN/spnf.aspx

Sulok, K. M. T., Ahmed, O. H., Khew, C. Y. and Zehnder, J. A. M. (2018). Introducing Natural Farming in Black Pepper (Piper nigrum L.) Cultivation. International Journal of Agronomy. **1**: 6. DOI:10.1155/2018/9312537.

Tripathi, S., Tauseef, S., Nagbhushan, S. and Gupta, N. (2018). Zero Budget Natural Farming for the Sustainable Development Goals, Andhra Pradesh, India.

2

Role of the Stakeholders in Agricultural Innovation System

P.H. Vihariya[1], K.V. Gardhariya[2] and R.D. Pandya[3]

[1&2]Extension Education Institute, AAU, Anand, Gujarat
[3]NMCA, Navsari Agricultural University, Navsari, Gujarat

Abstract

Role is a set of connected behaviours, rights, obligations, beliefs and norms as conceptualized by an individual in a social situation. Role is the task performed by an individual is expected to play in AIS. It is a set of socially expected and approved behaviour patterns consisting of both duties and privileges, associated with a particular position in AIS. This infers that the AIS generate a lot of hope for innovative technologies and interactive learning among stakeholders to overcome the weak linkages among them. The present study was carried out in seven districts of south Gujarat during 2020. The six types of stakeholder were playing pivotal role in AIS hence the list of each type of stakeholders were obtained from the concerned authorities. A simple random sampling method was adopted to obtain respondents sample size. Thus, 30 researches, 50 extensionists, 30 In –charge of NGOs, 30 managers of private agencies, 50 owner of agro-service providers and 50 progressive farmers were selected. All the 240 stakeholders were randomly selected. The statistical tools and method was used to analyse the data were frequency, percentage and arbitrary method. It can be concluded that more than two fifth (46.67 and 45.83%) each of the stakeholders performed their good and excellent role in AIS, followed by 7.50 per cent of them execute their poor role in AIS. Thus, it can be said that overwhelming majority (92.50%) of the respondents as stakeholders executed their good to excellent role in AIS.

Keywords: *Role, Stakeholder, Agricultural, Innovation System, Behaviour*

Introduction

In one or another way, agriculture is integral to the physical and economic survival of every human being. The United Nations forecasts that the global population will reach more than 9 billion by 2050. To feed everyone, food production will have to increase 70 per cent. Helping the world's farmers to achieve this target is challenging in itself, but beyond providing food, agriculture sustains the economies of most countries in significant ways, especially in the developing world (Anonymous, 2018). To meet this challenge in a fast-changing world, new tools, processes and ways of organizing-*i.e.* innovations-are needed to make agriculture more efficient in economic, social and environmental terms (Hall, Mytelka and Oyeyinka, 2006). Past efforts have tended to focus solely on technology. Previous approaches to technological change in agriculture have been based around supplying technology in a simplistic, top-down manner. These approaches had some success for relatively simple innovations in conducive settings. However, these simplistic approaches are not adequate for the task of meeting the complicated challenges facing agriculture today. Historically, new agricultural technologies have been developed by public research organizations and then transferred to farmers by public extension services (Sulaiman, 2015). The emerging issue is that the widespread adoption of new agricultural technologies and practices is a challenge that often requires more than knowledge transfer to increase the economic, social and environmental value created through agriculture. Dealing with these multiple limitations of previous approaches and for facing future challenges, there was a need for involvement of stakeholders as well as research and technology development in coordinated manner. So, the emphasis of AEAS (Agricultural Extension Advisory Services) has shifted from technology transfer to improving the innovation capacity among different stakeholders. Thus, AIS focus on how innovation can be scaled up and innovation capacity can be developed among stakeholders.

In this context, the concept of Agricultural Innovation System (AIS) approach offers a new framework for analysing the role of science and technology and their interaction with other actors to generate goods and services. It recognizes the importance of certain types of relationships and linkages for information flow in ever-changing biophysical and social environments. If sustainable innovations are targeted, the key challenges faced by agricultural sector are food security and climate change aspects should be promoted. Stakeholders performed their activities in AIS with qualities and in quantity without wasting time and resources. They offer their products / services on one's desire or need. In this regards, an attempt was carried out to measure the **"Role of the stakeholders in Agricultural Innovation System"**

Research methodology

The present study was conducted in Gujarat state. All seven districts of South Gujarat *viz.,* Navsari, Valsad, Surat, Tapi, Narmada, Bharuch and the Dangs were selected. The AIS has been playing their roles with six types of stakeholders. They are Researchers of university, Extensionist of line Departments, In-charge of agricultural oriented Non-Government Organization, Manager of agricultural oriented private agencies, Agro-Service Providers and Progressive farmers. These six type of stakeholder were playing pivotal role in AIS hence their list of each type of stakeholder were obtained from the concerned authorities. The mode of study area of research force to work was vertically as well as horizontally therefore; a simple random sampling method was adopted to obtain respondents sample size. Thus, 30 researchers, 50 extensionist, 30 In-charge of NGOs, 30 managers of private agencies, 50 owner of agro-service providers and 50 progressive farmers were selected. All the 240 stakeholders were randomly selected for present study.A simple understanding about role is the function assumed or part played by a person or thing in a particular situation. There have been six types of stakeholders in AIS and each has different roles. A structured scheduled was developed in context to AIS for measuring the identified role performed by the eachstakeholder.

The Researcher as stakeholder contains 12 roles, Extensionist contains 12 roles, NGO contains 11 roles, Manager of private agency contains 14 roles, Agro-service provider contains 7 roles and Progressive farmer contains 9 roles in AIS. The responses of stakeholders about their role were recorded on three point continuum which were frequently, sometime and never with 2, 1 and zero score respectively. An arbitrary method was used for categorisation to each section as well as for pooled data. For that the higher score is subtracted from the lower score and divided by the number of categories. The obtained score is added into the lower score until you get the highest score. Moreover the figures in decimal were round up in this case.

Responses regarding role of each stakeholder in AIS were collected through structured schedule and grouped into three categories. The categories for researchers and extensionists were as, (i) poor level of role (up to 8 score), (ii)good level of role (8 to 16 score) and (iii) excellent level of role (above 16 score), for In-charge of NGOs the categories were as, (i) poor level of role (up to 7 score), (ii) good level of role (8 to 15 score) and (iii) excellent level of role (above 15 score), for manager of private agencies the categories were as, (i) poor level of role (up to 9 score), (ii) good level of role (10 to 19 score) and (iii) excellent level of role (above 19 score),for agro service providers the categories were as, (i) poor level of role (up to 5 score), (ii) good level of role (6 to 9 score) and (iii) excellent level of role (above 9 score) and for progressive farmers the categories were as, (i) poor level of role (up to 6score), (ii) good level of role (7 to 12 score) and (iii) excellent level of role (above 12 score). The classified data are presented in below this table.

Table 1: Classified Data.

S.No.	Categories	Class range (Score)					
		Rese-archer	**Extio-nist**	**NGO**	**Manager Pvt.**	**Agro Ser. Prov.**	**Prog. farmer**
1	Poor role	0 to 8	0 to 8	0 to 7	0 to 9	0 to 5	0 to 6
2	Good role	8 to 16	8 to 16	8 to 15	10 to 19	6 to 9	7 to 12
3	Best role	17 to 24	17 to 24	16 to 22	20 to 28	10 to 14	13 to 18

Results and Discussion

The result presented in table 1 indicates that majority (86.67 %) of the researchers executes their excellent level of role and 13.33 per cent of them performed their good level of role in AIS. Same table shows that majority (64.00%) of the extensionists performed their excellent role and 36.00 per cent executed their good role in AIS. Whereas, majority (63.33%) of the In-charge of NGOs performed their excellent role and 36.67 per cent of them executed their good role in AIS.

Table 2: Distribution of respondents according to their role in AIS

Sr.	Type of Stakeholders	Poor Role	Good Role	Excellent Role	Total
1	Researcher (n=30)	00 (00.00)	04 (13.33)	26 (86.67)	30 (100.00)
2	Extensionists (n=50)	00 (00.00)	18 (36.00)	32 (64.00)	50 (100.00)
3	NGOs (n=30)	00 (00.00)	11 (36.67)	19 (63.33)	30 (100.00)
4	Private Agencies (n=30)	00 (00.00)	09 (30.00)	21 (70.00)	30 (100.00)
5	Agro-service provider (n=50)	18 (36.00)	32 (64.00)	00 (00.00)	50 (100.00)
6	Progressive farmers (n=50)	00 (00.00)	38 (76.00)	12 (24.00)	50 (100.00)
Pooled (n=240)		18 (07.50)	112 (46.67)	110 (45.83)	240 (100.00)

(Figures in parentheses indicate percentage to total)

Further, majority (70.00%) of the manager of private agencies performed excellent role and 30.00 per cent of them executed their good role in AIS. The majority (64.00%) of the agro-service providers performed their good role and 36.00 per cent executed their poor role in AIS. Moreover, majority (76.00%) of the progressive farmers performed their good role and 24.00 per cent of them executed their excellent role in AIS.

The pooled data shows that more than two fifth (46.67 and 45.83%) each of the stakeholders performed their good and excellent role in AIS, followed by only 7.50 per cent of them execute their poor role in AIS.

Thus, from the above result it can be concluded that overwhelming majority (92.50%) of the respondents as stakeholders executed their good to excellent role in AIS. The probable reason might be that the stakeholders they had high level of innovativeness, cohesiveness, their annual income, information seeking behaviour, extension participation, communication skill and decision making ability. The distinctive feature of AIS has the involvement of multi-stakeholders at different levels. All the stakeholders who innovate and share innovation, provide demands to AEAS, agricultural research system and interact with other stakeholders through exchange of information, inputs/goods and services. Private sector actors (agro processors, input dealers, distributors, traders, corporate houses, *etc.*) mostly innovate and share innovation, assess demand, provide and distribute inputs, purchase, process, trade, and export produce, educate and advise.

Conclusion

As the study showed that majority of the researchers, extensionists, In-charge of NGOs, manager of private agencies, agro-service providers and progressive farmers executes their good to excellent level of role in AIS.

Much of the research has not been conducted on role of stakeholders in Agricultural innovation system, so this was a maiden and novel investigation in Gujarat state, which tries to analyze the role of stakeholders perform by them in Agricultural innovation System. Their role performance leads to quality work and the study showed that stakeholders had really perform excellent in AIS which made AIS more productive.

References

Anonymous, (2018). www.fao.org/india/fao-in-india/india-at a-glance/en/

FAO. 2017. Enabling the Capacity to innovate with a system-wide assessment, process, Food and Agriculture Organization of the United Nations Rome.

Gardhariya, K. V.; Pandya, R. D.; Bhuva, R. M. and Dhodia, A. J. (2016). Role perception and role performance of farm school beneficiary working under ATMA of south Gujarat. International Journal of Agriculture Science, 8 (58): 3197-3198.

Hall, A., Mytelka, L. and Oyeyinka, B. 2006. Agricultural Innovation Systems: A methodology for diagnostic assessments. In: Enhancing agricultural innovation: How to go beyond the strengthening of research systems, World Bank, Washington DC.

Klerkx, L. and Gilemacher, P. (2012). The role of innovation brokers in Agricultural Innovation Systems. In: Agriculture and Rural Development: An Investment Sourcebook. 221–230, Washington DC, World Bank.

Suchiradipta, B. and Saravanan R. (2014). Agricultural Innovation Systems (AIS): A Study of Stakeholders and their Relations in System of Rice Intensification (SRI), The J. of Agri. Edu. And Ext., DOI: 10.1080/1389224X.2014.939200

Sulaiman, R.V. (2015). Agricultural Innovation Systems. Note 13. GFRAS Good Practice Notes for Extension and Advisory Services. GFRAS: Lindau, Switzerland.

World Bank. (2012). Agricultural Innovation Systems: An Investment Sourcebook. Washington DC, World Bank. Retrieved from http://siteresources. worldbank.org/INTARD/ Resources/3358071330 620492317/9780821386842.pdf

3

Extension Approaches for Climate Smart Agriculture

***Souvik Ghosh*[1] *and Usha Das*[2]**

[1]*Agricultural Extension, Palli Siksha Bhavana (Institute of Agriculture) Visva-Bharati University, Sriniketan, Birbhum, West Bengal*
[2]*National Institute of Agricultural Extension Management Rajendranagar, Hyderabad, Telangana*

Developing countries with agrarian economy have been facing challenges arising from climate change causes and consequences. The altered patterns of climate are affecting the yield potential of crops thereby affecting agricultural production, farmers' income, and livelihoods. Out of the 17 Sustainable Development Goals (SDGs) of the United Nations, the 13th SDG on 'Climate action to combat climate change and its impacts,' emphasises the importance and urgency of addressing the effects of climate change in developing nations. It is projected that climate change would have the most detrimental effect on agriculture, the primary source of income for small and marginal farmers in developed nations (Azadi et al., 2021). Building the ability of small and marginal farmers will be crucial to global food security over the next decade (FAO, 2020). World Bank (2018) reported that 800 million South Asian population are at risk of experiencing a decline in their standard of living and incomes due to rising temperatures and more erratic rainfall hampering the agriculture production system. Millions of low-income smallholder farmers, particularly those who rely on agriculture for a living and an income in South Asia, would be directly impacted by a reduction in productivity leading to food supply shortages and price increases.

Climate Smart Agriculture (CSA) is being promoted to combat the climate change issues and their effect on agriculture. CSA can transform the current agricultural system to support food security and safeguard it during increased climate vulnerabilities; therefore, need of present-day agriculture is to transform into CSA establishing a sustainable agricultural production system and stable food-secured system (Das et al., 2022a). Farmers have differential awareness, knowledge, and perceptions towards climate change and thus their responses differ to combat climate change impacts (Karki et al., 2020; Ramborun et al., 2020). Climate change adaptation strategies are being formulated in the agriculture sector;

however, consistency of those with the farmers' preferences has been the concerns in developing countries (Khanal et al., 2018). Climate change mitigation measures are often hampered by farmers' awareness, knowledge and attitude towards climate change and its impact on agriculture & vice versa (Tzemi and Breen, 2019). Thus, the need Climate Smart Extension (CSE) approaches have been aroused in recent years.

Indian Agriculture in Changing Climate Scenario

Climate change has emerged as a serious threat to the livelihood of Indian farmers with about 82 percent of farmers from small and marginal category (FAO, 2020), who are still poorly informed and inadequately prepared to face the changing climate scenario and its impacts on agriculture. Changes in the monsoon (time and intensities), an increase in the frequency and severity of severe storms, flood, and drought have impacted Indian agriculture and agrarian livelihoods. Indian Council of Agricultural Research (ICAR, 2015) in 'Vision 2050' considers climate change as a major threat to long-term food security as it may have multidimensional negative effects on farming sector.

Due to limited capacities, inadequate institutional systems, and lack of access to necessary resources, the majority of India's vulnerable population is ill-equipped to adequately deal with the effects of climate change (Kumar, 2008). Climate change has brought widespread misery and huge economic losses to India, adversely affecting agriculture, food security, public health, water resources and biodiversity (Sarkar et al., 2014). Das and Tripathi (2014) reported that 40% farmers want to quit farming. In India, the vulnerability of farming community towards climate change is more due to its poor socio-economic status, poor infrastructural base, poor cognitive empowerment, and huge technological gap at farm level. Hence, vulnerability of farming community extends beyond bio-physical impact and includes multiple socio-political, economical, and psychological aspects. With a specific staple crop dominant cropping system in most regions of India like Rice dominant cropping systems in Eastern India, Wheat and Maize dominant cropping systems in Northern region, these areas are more vulnerable to climate change than the areas with diverse cropping systems.

The issue of climate change in agriculture has been viewed more from the perspectives of physical, chemical, and biological sciences research (Ghosh at al., 2017). Most of these researches were on forecasting, modelling, simulation studies, mitigation of greenhouse gases, energy and emissions, use of renewable energy, climate resilient farming interventions through resistant varietal development (evolving plants that use water more efficiently, resist drought, flood, pests, etc.), natural resource management (water conservation measures, soil fertility, etc.), bio-resource conservation, etc. The contribution of social science in general and Agricultural Extension along with EAS in particular has been largely restricted to impact assessments, capacity building and demonstrations through Krishi Vigyan Kendras (KVKs). However, there is a need of paradigm shift to address the issues of resilience of agriculture, farm livelihood assets, farmers' climate knowledge, livelihood sensitivity, social and economic vulnerability, and household as well as community level resilience, and promotion of CSA interventions (both technical and institutional) through participatory prioritisation and scaling up in climatically vulnerable regions (Fig. 1).

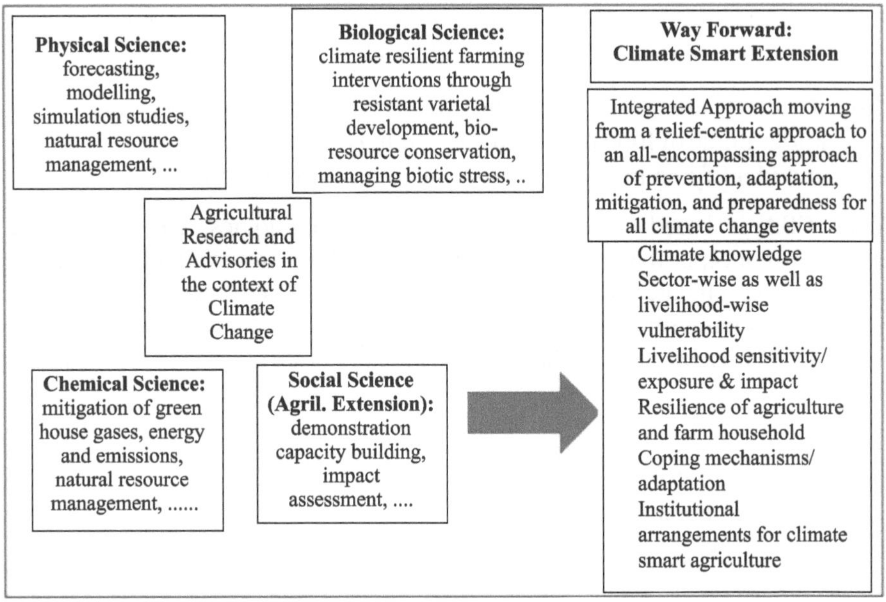

Fig. 1: Paradigm shift in extension approaches to addressing climate change issues in agriculture

Climate Smart Agriculture

Climate Smart Agriculture (CSA) integrates the climate change issues into the planning and development of sustainable agricultural production systems. The concept of CSA is given by Food and Agricultural Organization of the United Nations (FAO 2013). CSA is defined as "agriculture that sustainably increases productivity, enhances resilience (adaptation), reduces/removes GHGs (mitigation) where possible, and enhances achievement of national food security and development goals". According to FAO (2013), there are three pillars of CSA- (i) increasing agricultural productivity to support sustainable livelihoods (ii) adaptation and capacity building to bring out resilient agricultural systems (iii) mitigation by reducing the GHGs emissions to a safe level. The domains of CSA are food security, adaptation, and mitigation (Fig. 2).

CSA is an integrated approach for managing agricultural production system (crops, livestock, forests, and fisheries) that addresses the interdependent issues of food security and increasing climatic change. CSA seeks to concurrently attain triple wins:

i. Increased productivity: Sustainable food production ensuring food and nutrition security and increased incomes of rural people relying mostly on agriculture for a living.

ii. Enhanced resilience: Improving capacity and adaptation to sustain in the face of climatic aberrations like shortened seasons, erratic weather patterns, etc and reduced vulnerability to flood, drought, pests, diseases and other climate-related risks and shocks.

iii. Reduced emissions: Striving for lower emissions per calorie or kilogramme of food produced, preventing deforestation from agriculture, and identifying methods to absorb carbon from the atmosphere.

While CSA is based on current knowledge, technologies, and sustainable agriculture concepts, it is distinct in several aspects. First, it is explicitly committed to combating climate change. Secondly, CSA evaluates systematically the synergies and trade-offs between productivity, adaptation, and mitigation. Lastly, CSA intends to seize new funding opportunities in order to close the investment gap.

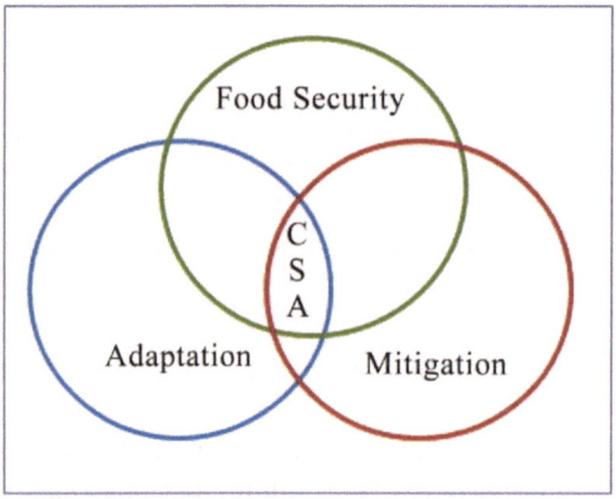

Figure 2: Domains of climate smart agriculture (CSA)

CSA encourages coordinated actions by farmers, researchers, the private sector, civil society, and policymakers toward climate-resilient pathways through four main action areas: (1) building evidence, (2) increasing local institutional effectiveness, (3) fostering coherence between climate and agricultural policies, and (4) linking climate and agricultural financing. Thus, CSA varies from business-as-usual approaches as it emphasises the capacity to deliver flexible, context-specific solutions supported by creative policy and finance initiatives (Lipper et al., 2014).

Climate Smart Extension

Climate smart extension (CSE) is an advisory services for transferring various forms of climate information to farmers in order to help them to make informed farming decisions, implement climate-smart technologies to adapt to climate change and mitigate its effects, and reduce GHG emissions in agricultural practises in order to achieve global food security (Prasanth and Palaniappan, 2019). EAS may play a vital role in promoting CSA by transmitting real-time climate information to farmers. It helps to increasing farmers' awareness and knowledge of climate change. Various extension methods, which can be used effectively to strengthen the capacity of farmers to adopt CSA (Singh, 2020) are:

- Climate awareness programmes/ campaigns, exhibition
- Climate workshops
- Climate trainings
- Climate smart farmers field schools (FFS)
- Field visits to progressive farmers
- Demonstration on different adaptation/mitigation practices
- Dissemination of appropriate climate resilient technology
- Contingency crop planning
- *Jaldoot* - community level extension professional
- Community based disaster management (CBDM) approach
- Village climate risk management committee (VCRMC)
- Agro-meteorological advisory service
- Village level custom hiring centres (CHCs)

Extension and advisory services (EAS) can play a crucial role in upscaling and out scaling of CSA. EAS contribute to the achievement of all three goals of the CSA (food security, adaptation, and mitigation), although their current emphasis is on food security through increased productivity. EAS must now be implemented more aggressively to assist rural populations in adapting to climate change and to contribute to climate change mitigation (Sulaiman et al., 2017).

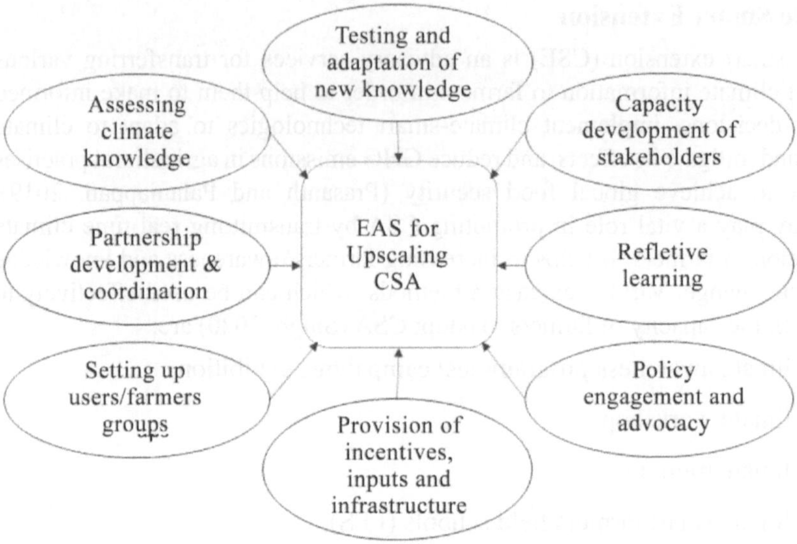

Figure 3: Functions of EAS to upscaling CSA interventions (Source: Sulaiman et al., 2018)

EAS need to include an array of function being an integral part of agricultural innovation systems (Fig. 3). Upscaling of CSA is not only merely disseminating information or advising farmers on how to adopt a climate smart variety, practice, or product, which is a necessary but not sufficient precondition to achieving desirable impact of CSA, and therefore must be combined with other innovation management functions that can be accomplished through building partnerships. EAS must engage in testing and adaptation of new knowledge, reflective learning, provisioning of incentives, inputs & infrastructure, setting up user/ client groups, capacity development of all stakeholders, and policy engagement & advocacy (Sulaiman et al., 2018). Identification and upscaling of CSA practices will need generation of knowledge in ever changing climate scenario. Monitoring, evaluation and learning mechanisms will be part of knowledge management strategy of EAS systems. If EAS providers are to contribute significantly in screening and upscaling CSA they also require to engaging more in research.

Integrated Approach to Promote CSA through CSE

Extension education, research as well as advisory services need to focus on the agriculture, water resources/ irrigation, climatic vulnerability, and socio-economic scenario of a climatically vulnerable region to screen, prioritise and promote location specific CSA interventions. The scenario analyses may like to consider time series data on various parameters over the space and time to identify climatically vulnerable as well as resilient regions with the help of different indexes. Depending on the availability of data, scenario analyses may be undertaken at the grassroot level starting from the district level.

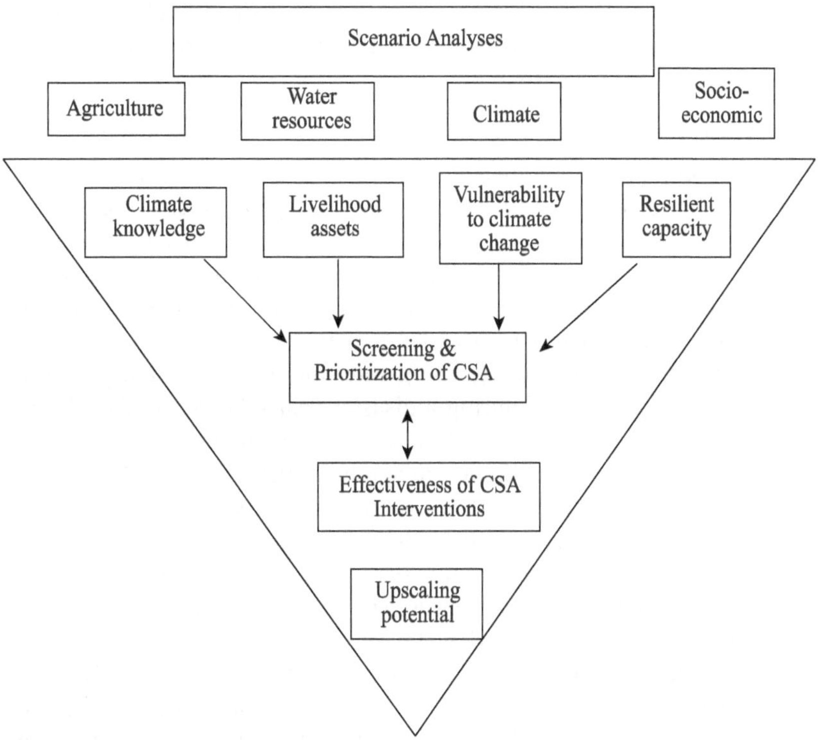

Figure 4: Extension research and advisory services to promote CSA

Following the scenario analyses, farm household level analyses will be important with respect to the farmers' climate knowledge including the causes and consequences of the climate change, livelihood status as determined by physical, social, human, financial, and natural assets holdings of farm households, farmers' vulnerability to climate change in terms of exposures to various climatic events (cyclone, flood, drought, heatwave, etc.) and their impacts, and resilient capacity of farm households depending on their access to basic services, assets, social safety nets and adaptive capacity. These farm households' attributes act as the determinants of adoption of CSA in a region.

An inventory of CSA interventions in terms of scientific technologies, indigenous technical knowledge, information and communication technologies, and institutional innovations will initiate the screening and prioritization process through assessment of effectiveness of the CSA interventions and their upscaling potentials. The involvement of farmers in the assessment process will take care farmers' preferences influencing the better adoption of CSA interventions. An integrated approach of extension education, research and advisory services will be able to design and deploy climate smart socio-technical innovations bundling in terms of both technological and institutional innovations through an effective partnership of all stakeholders, which will provide diversified livelihood options to the farm households and better insulation against the vulnerability to climate change in a specific agro-ecosystem.

Scenario Analysis for CSA & CSE

Promotion of appropriate location-specific CSA interventions require an analysis of scenarios of agriculture, water resources for irrigation, socio-economy and climatic vulnerability of the regions. It would reflect the resilience of the region's agriculture to the climate change as well as climate induced natural disasters. In a study, a composite agricultural scenario index (ASI) was constructed considering 10 years data on 11 parameters to depict the district-wise agricultural scenario in West Bengal (Ghosh et al., 2017), The greater the ASI value, the greater the resilience exhibited by a district during a disaster year relative to other districts that experienced the same occurrence. As a result, the agricultural scenario of six districts in North Bengal and five districts in South Bengal suffered the most during flood years, as evidenced by the lowest ASI values for those years. As demonstrated by the lowest ASI value, districts in the state's western region suffered the most from the drought. Resilience of agriculture depends not only on the sustainability of agricultural-related parameters as measured by ASI, but also on socio-economic condition, climatic vulnerability and water resources for irrigation; therefore, the socio-economic index (SEI), season-wise climaticvulnerabilityindex(CVI), groundwater development index (GWDI), surface irrigation index (SII), and overall irrigation scenario index (ISI) wereworked out (Ghosh et al., 2017a and 2017b).

Using Arc GIS Desktop Basic Software, index maps are generated to illustrate the comparative agricultural, socioeconomic, climatic vulnerability, and water resources/ irrigation scenario of West Bengal's districts. There is no link between agriculture and the socio-economic scenario; however, it is evident that the socio-economic scenario is at a medium level in most districts (Fig. 5). Thus, the socio-economic situation of districts has not yet been influenced by agricultural performance, which makes the impact of climatic vulnerability on agricultural production systems, mostly operated by marginal and small farmers, even more significant. It becomes more difficult to improve the socio-economic situation as climate-related hazards increase. Fig. 6 demonstrates that pre-monsoon vulnerability exists at a very high level in two districts during the summer months, while seven districts (6 districts in South Bengal and the Cooch Behar district in North Bengal) are highly vulnerable during monsoon season. During the post-monsoon season, coastal districts such as East Midnapore and South 24 Parganas are most susceptible to cyclonic storms originating from low depressions in the Bay of Bengal. There is also vulnerability during the winter season, but it is more dispersed across districts and regions in West Bengal. West Bengal's three most vulnerable districts to climate change are South 24 Parganas, Cooch Behar, and

Purulia, when considering the ASI, SEI, and CVI together. Looking at the water resources scenario for irrigation in terms of GWDI, SII and ISI (Fig. 7), it is evident that groundwater irrigation is predominant in Gangetic plains of South Bengal, while surface water is the prime source of irrigation in rest of the South Bengal districts. Composite irrigation scenario index (ISI) depicts the lack of irrigation facilities in North Bengal districts. The performance of agriculture is factored by irrigation availability; therefore, irrigation availability develops better resilience to agriculture in the event of climatic aberrations.

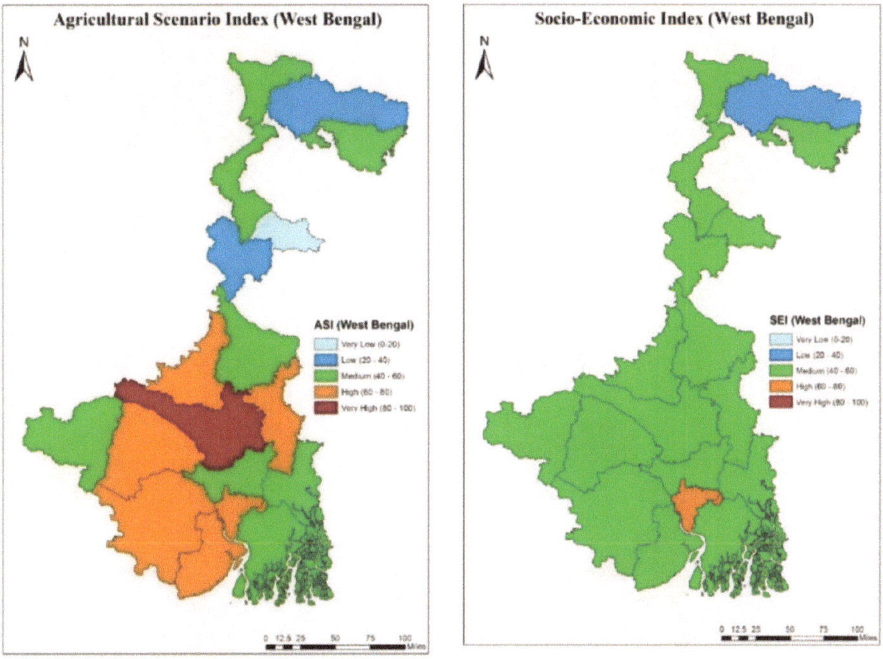

Figure 5: Agricultural and socio-economic scenario depicted through ASI and SEI in districts of West Bengal (Source: Ghosh et al., 2017a)

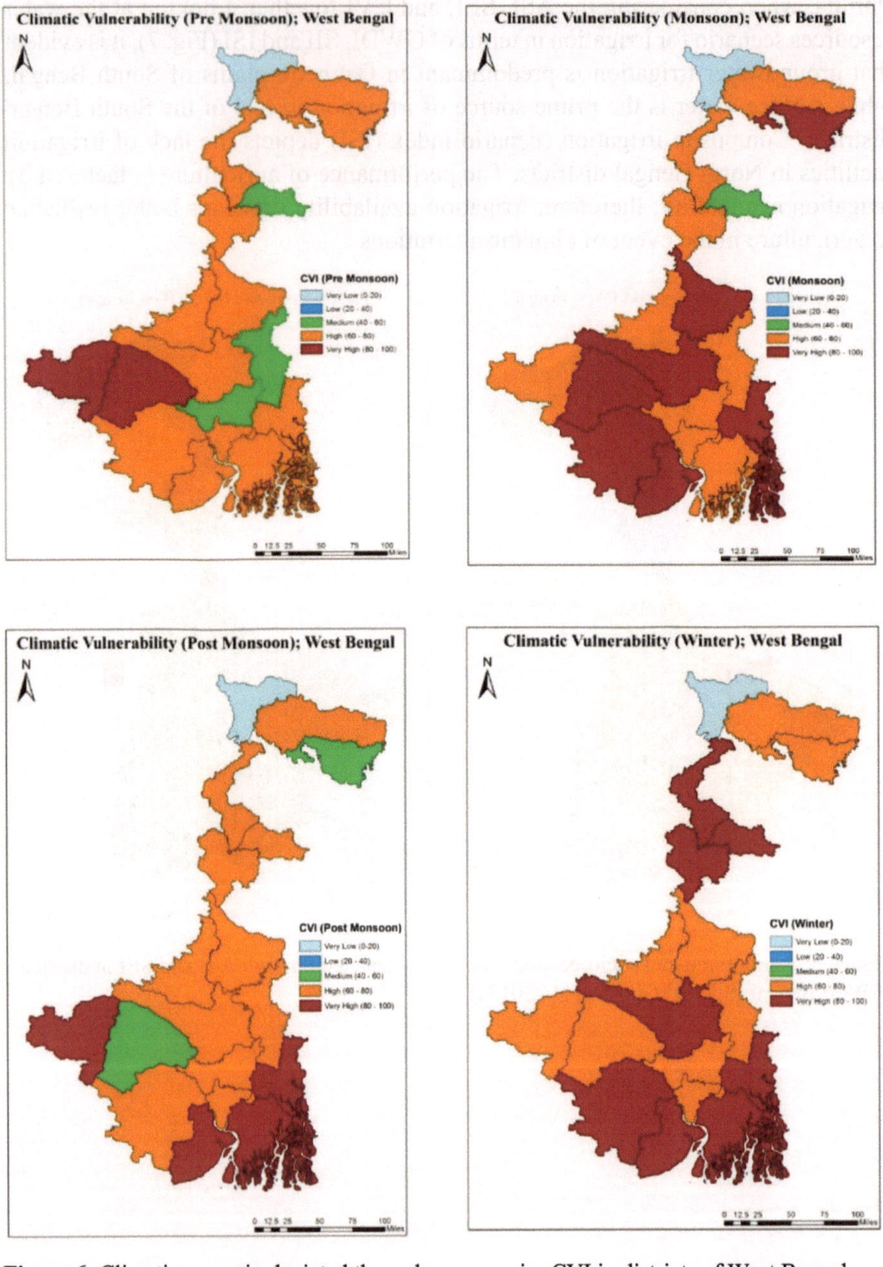

Figure 6: Climatic scenario depicted through season-wise CVI in districts of West Bengal (**Source:** Ghosh et al., 2017a)

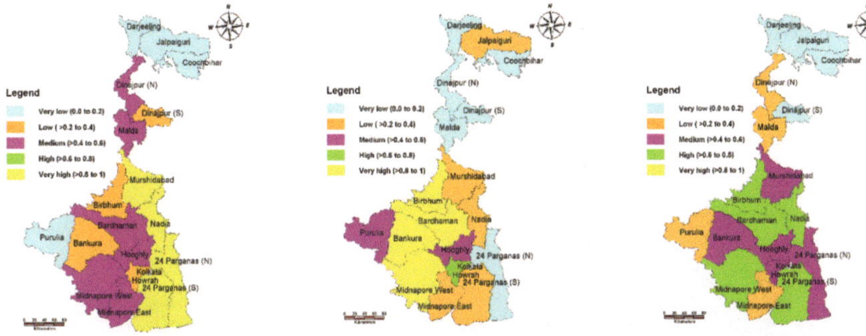

Figure 7: Groundwater, surface water and overall irrigation scenario depicted through GWDI, SII and ISI in districts of West Bengal (Source: Ghosh et al., 2017b)

Similarly, another study (Das et al., 2020) found that the maximum and minimum ASI values in all 30 districts of Odisha fluctuated due to climate anomalies. In comparison to other districts, the coastal districts of Balasore and Bhadrak have demonstrated a greater resilience of agriculture during years of extreme climatic events, keeping a higher ASI value. This is a result of improved groundwater irrigation and socioeconomic conditions in these districts. However, other coastal districts have demonstrated less resilience, with a higher decline in ASI value during years of climate-related hazards.

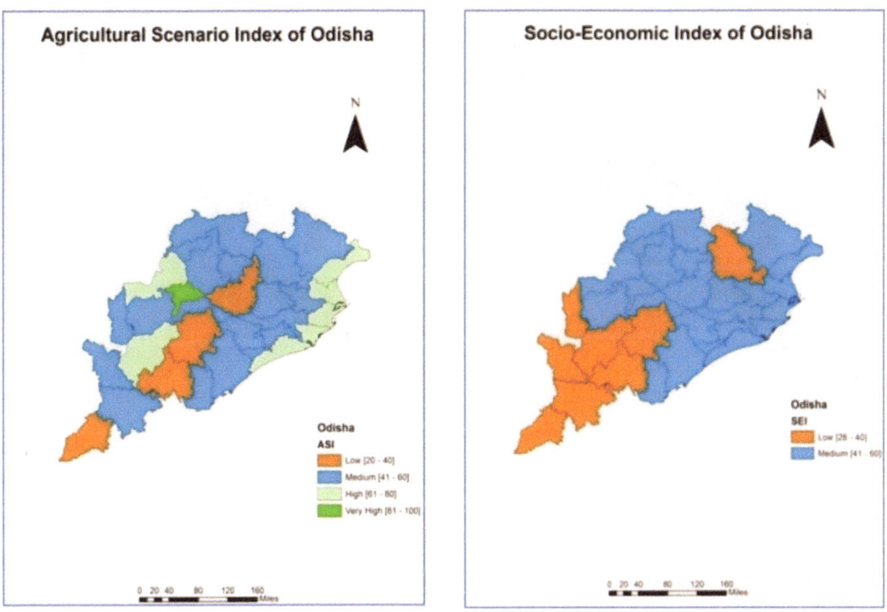

Figure 8: Agricultural and socio-economic scenario depicted through ASI and SEI in districts of Odisha (Source: Das et al., 2020)

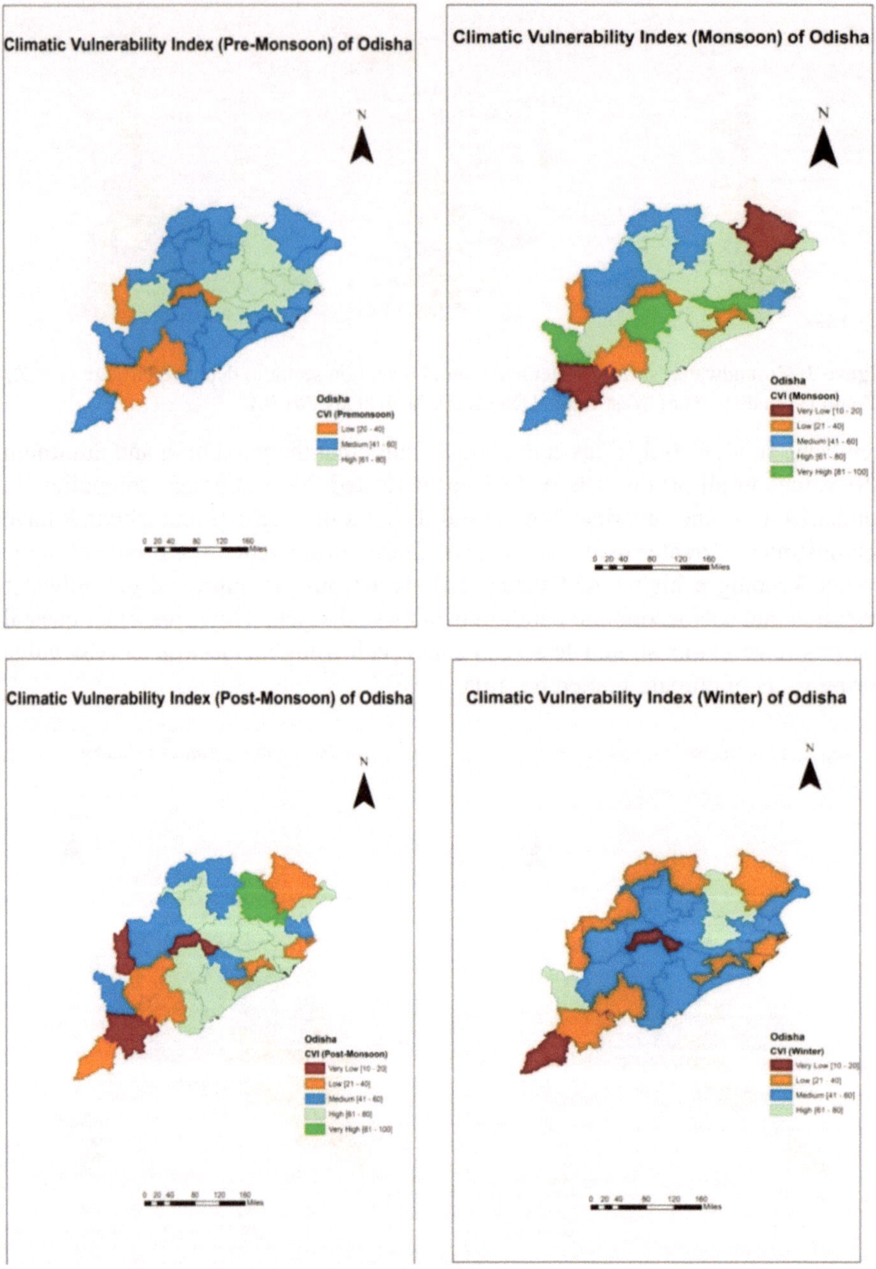

Figure 9: Climatic scenario depicted through season-wise CVI in districts of Odisha (Source: Das et al., 2020)

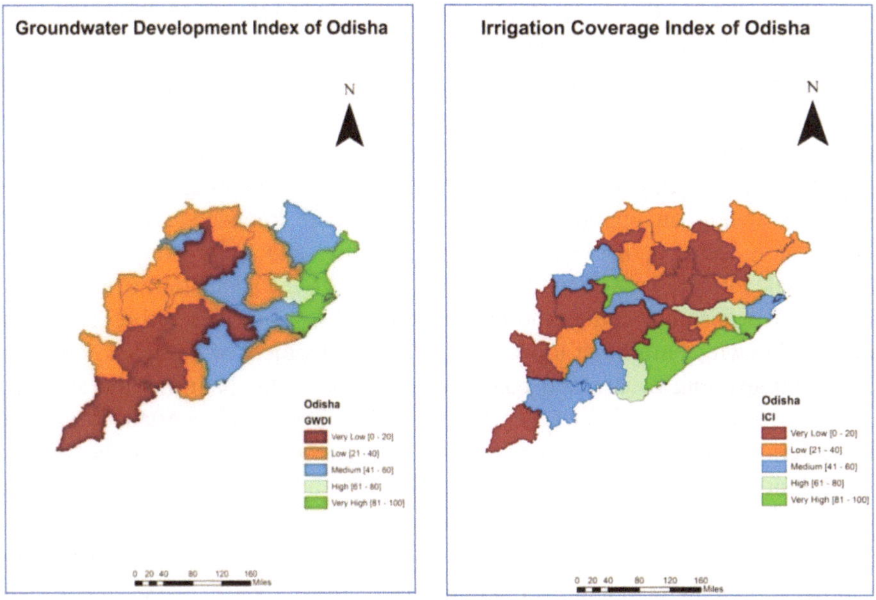

Figure 10: Groundwater and irrigation coverage depicted through GWDI and ICI in districts of Odisha (Source: Das et al., 2020)

Evidently from Fig. 8, Fig. 9 and Fig. 10, the districts of Odisha have showed both links and missing links between agriculture, water, climate, and socio-economic scenarios. Link is evident between agriculture and irrigation; however, the same is not evident between agricultural and socio-economic scenario. Seasonal climatic vulnerability index depicts relatively more vulnerability of coastal districts during monsoon and post-monsoon seasons. socio-economic scenario is found at medium level in most districts. With increasing climatic vulnerability, it becomes more challenging to improve upon the socio-economic condition to increase the resilience of farm households towards climate change events. Future efforts for climate resilient agriculture need to be directed to less resilient districts to bettering income of farmers. Future climate smart agriculture efforts must be focused on less resilient regions in order to reduce the susceptibility of farming livelihoods to climate change.

Farmers' Participatory Prioritization of CSA and Its Drivers

The major challenge for CSA addressing location specific climate change issues is identification, prioritization, and upscaling of available CSA interventions in diverse agro-ecosystems. The decisions made by policymakers often fail to be realistic and functional without listening to the voices of the farmers (Azadi et al., 2021). The challenges to promote adaptation to climate change can be met through involvement of stakeholders, where opinion of farmers, the primary stakeholder is of pivotal importance. As adaptation bears a better observability compared to mitigation, it becomes more alluring for small scale farmers to adopt the CSA practices underlining adaptation.

Enhancing the need and capacity to manage climate risk, which is a core adaptation strategy, necessitates a rise in the climate knowledge of decision makers so that they are more aware of climate impacts on various systems and how to use management options to intervene, thereby mitigating negative impacts and capitalising on opportunities. Lack of climate knowledge in the society creates a wider gap between policy formulation and policy implementation. Inadequate climate knowledge can not only weaken the adaptation strategies but it may also misguide mitigation efforts in this dimension. Therefore, assessing climate knowledge through location specific development of climate knowledge test encompassing causes and consequences of climate change is the stepping stone in providing EAS for CSA (Das et al., 2022b). It is important to unravel the determinants of climate knowledge of the stakeholders including the farmers that would be precursor to formulate location-specific as well as clientele-specific capacity building programmes and EAS (Das and Ghosh, 2020).

The prioritization of CSA interventions by the farmers as adaptation options following an integrated approach will enhance the acceptance of those; one such location-specific framework to consider farmers' inputs in prioritization and upscaling of CSA is suggested that consider productivity, income, resilience and mitigation to determine effectiveness of CSA practices, and technical feasibility, cost of technology, gender inclusivity, and synergy with government plans as the parameters to determine upscaling potential of CSA interventions (Das et al., 2022a). Delineation of livelihood indicators as drivers of CSA would help in policy advocacy to promote CSA. Government support in terms of financial assistance, subsidies, and insurance, climate smart EAS, adequate institutional infrastructure, and effective partnerships between the stakeholders would enable farmers to adopting CSA innovations, building sustainable farm livelihoods and improving resilience to climatic change.

Progress of CSA and CSE in India

National Innovations on Climate Resilient Agriculture (NICRA) was launched by ICAR in 2011 to develop different climate resilient technologies and transfer the same to different vulnerable districts of India. Several leading projects under The CGIAR Research Program on Climate Change, Agriculture and Food Security (CCAFS) are continuing, one of them being Climate Smart Villages (CSVs). Climate Change Knowledge Network in Indian Agriculture (CCKN-IA) is another initiative of MoA&FW launched in 2013 that is focused on ICT enabled approaches for knowledge exchange on climate change in Indian context. ICAR has implemented National Innovations on Climate Resilient Agriculture (NICRA) from 2011 to promote various resilient technologies to different vulnerable districts of India. India's take on climate change and climate smart agriculture has been encouraging but not sufficient. Despite initiatives of national and international agencies to promote CSA in India, there is still low adoption of such interventions (Aryal et al., 2018). The adaptation strategies remain unique to a particular setting and blanket suggestion for its up-scaling and out-scaling has remained an issue even decade after the introduction of NICRA. Therefore, the climate change adaptation policies must be integrated with climate smart extension and advisory services (EAS). Under NICRA of ICAR, 151 Climate Smart Villages are addressing vital challenges like drought, flood, soil related problems, etc. These climate smart villages are aimed to become: weather smart, water smart, carbon smart, nutrient smart, energy smart, and knowledge smart. The number and kind of interventions implemented are largely determined by the resources available, vulnerability status and involvement of communities. Village Climate Risk Management Committee (VCRMC) is constituted to manage both climate smart technical and institutional interventions in a climate smart village (Singh, 2020).

Conclusion

Scaling up CSA will undoubtedly necessitate altering the behaviour, methods, and agricultural practises of millions of agricultural producers, who must become more knowledgeable about the effects of climate change in order to adopt more climate-smart strategies. EAS have long functioned as a link between research and agriculture, providing farmers with information about new technologies. However, the effective expansion of CSA requires methods that extend far beyond the modification of farm-level agronomic techniques. In fact, it necessitates the identification and promotion of suitable techniques, technologies, and/or models (new, improved, or adapted) within enabling settings, and it must include constructive institutional structures, policies, and financial investments. Thus, climate smart extension needs to be pluralistic in nature meeting the multifarious needs of farmers in a climatically vulnerable region. Approaching climate change issues in an integrated manner would facilitate the links between sustainable development policy and emerging climate policy on resiliency and adaptation. Climate smart extension requires to focus on climate smart socio-technical innovations bundling through good partnerships to provide farm households a better resilience to climate change in a climatically vulnerable agro-ecosystem.

References

Aryal, J.P., Jat, M.L., Sapkota, T.B., Khatri-Chhetri, A., Kassie, M., Rahut, D.B., & Maharjan, S. (2018). Adoption of multiple climate smart agricultural practices in the Gangetic plains of Bihar, India. International Journal of Climate Change Strategies and Management, 10(3), 407-427.

Azadi, H., Moghaddam, S.M., Burkart, S., Mahmoudi, H., Van Passel, S., Kurban, A., & Lopez-Carr, D. (2021). Rethinking resilient agriculture: From climate-smart agriculture to vulnerable-smart agriculture. Journal of Cleaner Production, 319, 128602.

Das, S.K., & Tripathi, H. (2014). India's green revolution: Fact and fallacy. International Journal of Bio-resource and Stress Management, 5(1), 153-158.

Das, U., & Ansari, M.A. (2021). Climate Change-Sustainable Agriculture-Farm Livelihood Nexus: Contextualizing Climate Smart Agriculture. Climate Research, 84, 23-40.

Das, U., & Ghosh, S. (2020). Factors driving farmers' knowledge on climate change in a climatically vulnerable state of India. Natural Hazards, 102, 1419-1434.

Das, U., Ansari, M.A., & Ghosh, S. (2022a). Effectiveness and upscaling potential of climate smart agriculture interventions: Farmers' participatory prioritization and livelihood indicators as its determinants. Agricultural Systems, 203, 103515.

Das, U., Ansari, M.A., & Ghosh, S. (2022b). Does climate knowledge act as a shield for farm livelihoods? Empirical analysis from the coastal and non-coastal ecosystems of India. Theoretical and Applied Climatology, 150, 1627–1642.

Das, U., Ghosh, S., & Mondal, B. (2020). Resilience of Agriculture in a Climatically Vulnerable State of India. Theoretical and Applied Climatology, 139, 1513-1529.

FAO (2013). Climate Smart Agriculture. Source book. Food and Agriculture Organization of United Nations. Available at https:www.fao.org/3/a-i5188e.pdf

FAO (2017). Sustainable Development Goals. Food and Agriculture Organization of the United Nations. Retrieved from: http://www.fao.org/ sustainable-development-goals/goals/goal-13/en/

FAO (2020). Food and Agriculture Organization of the United Nations. (2020). http://www.fao.org/india/fao-in-india/india-at-a-glance/en/

Ghosh, S., Gorain, S., & Mondal, B. (2017b). Spatio-temporal variations and links between irrigation and agricultural development in an eastern Indian state. Irrigation and Drainage, 66, 784-796.

Ghosh, S., Mahato, K., Gorain, S., Das, U., & Mondal, B. (2017a). Resilience of agriculture reducing vulnerability to climate change in West Bengal. Current Advances in Agricultural Sciences, 9(2), 170-177.

Karki, S., Burton, P., & Mackey, B. (2020). The experiences and perceptions of farmers about the impacts of climate change and variability on crop production: A review. Climate and Development, 12(1), 80-95.

Khanal, U., Wilson, C., Lee, B., & Hoang, Viet-Ngu. (2018). Smallholder farmers' participation in climate change adaptation programmes: Understanding preferences in Nepal. Climate Policy, 18(7), 916-927.

Kumar, R. (2008). Climate Change and India: Impacts, Policy Responses and a Framework for EU-India Cooperation. Policy Department Economic and Scientific Policy. TERI-Europe, London.

Lipper, L., Thornton, P., Campbell, B.M., Baedeker, T., Braimoh, A., Bwalya, M., Caron, P., Cattaneo, A., Garrity, D., Henry, K., Hottle, R., Jackson, L., Jarvis, A., Kossam, F., Mann, W., McCarthy, N., Meybeck, A., Neufeldt, H., Remington, T., Sen, P.T., Sessa, R., Shula, R., Tibu, A., Torquebiau, E.F. (2014). Climate-smart agriculture for food security. Nature Climate Change, 4, 1068–1071.

Prasanth, A., &Palaniappan, B. (2019). Climate Smart Extension approaches for Sustainable Food Security. International Conference on Extension for Strengthening Agricultural Research and Research and Development (eSARD), December 2019.

Ramborun, V., Facknath, S., & Lalljee, B. (2020). Moving toward sustainable agriculture through a better understanding of farmer perceptions and attitudes to cope with climate change. The Journal of Agricultural Education and Extension, 26(1), 37-57.

Sarkar, S., Padaria, R. N., Vijayaragavan, K., Burman, R.R., Pathak, H., Kumar, P., & Jha, G. K. (2014). Assessing the socio-economic impacts of climate change in arid ecosystem of India. Range Management and Agroforestry, 35(2), 249-255.

Singh, A.K. (2020). Handbook of Agricultural Extension. New Delhi, Indian Council of Agricultural Research (ICAR).

Sulaiman, R.V. (2017). Enabling advisory services for climate-smart agriculture: Key elements to foster farmers' adoption of CSA practices. Policy Brief. Rome, Food and Agriculture Organization of the United Nations (FAO).

Sulaiman, R.V., Chuluunbaatar, D., & Vishnu, S. (2018). Upscaling climate smart agriculture-lessons for extension and advisory services. Occasional Papers on Innovation in Family Farming. Rome, Food and Agriculture Organization of the United Nations (FAO).

Tzemi D, & Breen J. (2019). Climate change and the agricultural sector in Ireland: examining farmer awareness and willingness to adopt new advisory mitigation tools. Climate Policy, 19(5), 611-622.

World Bank (2018).South Asia's hotspots: impacts of temperature and precipitation changes on living standards. World Bank, Washington, DC.

4

Climate Change Adaptation of Farming Communities in Western Ghat Region

Sreeram Vishnu[1], Archana Bhatt[2], Jeena Pilsy Paul[3], Denny Franco[4], Archana Raghavan Sathyan[5], Smitha Shiva[6] and Ajith Kumar[7]

[1,4,,5,6]*Kerala Agricultural University*
[2]*Community Agro Biodiversity Center, M S Swaminathan Research Foundation, Wayanad*
[3]*Independent Researcher*
[7]*Regional Agricultural Research Station, Ambalavayal, Kerala Agricultural University, Kerala*

Abstract

This study was conducted to assess the knowledge and extent of adoption of climate resilient farming practices by the growers in the Wayanad district of Kerala, falling under Western Ghat region. Various studies had already established vulnerability of the region to the impact of climate change and the resulting loss of crops and agro- biodiversity.Hence there is an urgent need to understand the adaptive responses of farmers to deal with such challenges. Data was collected from 251 practicing farmers from 6 villages, by following proportionate random sampling for the purpose of this study. Among the various adaptation practices, majority (>80 %) of the farmers practice land and soil management such as mulching, use of soil ameliorants as well as crop management practices like crop diversification, mixed cropping and agro- forestry. However, low level of adoption was observed regarding ICT based adaption measures. Further, many crop based adaptation measures such as irrigation scheduling, use of bio fertilizers and split dose of fertilizer application are not translated to adoption though a sizable proportion (>60 %) had knowledge about such practices. Results suggested that adoption of adaptation practices were correlated with farming experience, availability of subsidy for the technology, ready availability of resources within the farm and technical support for adoption of the practice. The study throws light into the level of adoption of climate adaptation practices by the farming community and vulnerable areas in which their adaption capacity has to be strengthened.

Keywords: *Western Ghats, Climate vulnerability, Climate resilience, Technology adoption, Capacity building*

1. Introduction

Agriculture has always been greatly dependent on climate patterns and variations, and is considered to be particularly at risk to the influences of climate change (James et al., 2014). Climate change is manifest in the form of reduction in rainfall and drought experienced in the last decade, pushing the agrarian community into real distress (MSSRF, 2020). To reduce the adverse impact of climate change on agriculture, adaptation is considered a crucial component of any policy response related with agriculture and climate change (Deressa et al., 2009). Adaptation to climate change involves changes in agricultural management practices in response to changes in climate conditions (Akinnagbe and Irohibe, 2014). Having knowledge about climatic adaptation practices by the farming communities would make them better equipped to deal with climate change related challenges (Mwazi&Ndokosho, 2011). Though smallholder cultivators are one of the most vulnerable groups to climate change, efforts to support their adaptation strategies are hindered by the lack of information on how they are experiencing and responding to climate change (Harvey et al., 2018). Senapati (2019) argues that farm households with access to weather-related information and extension agency contact have lesser level of vulnerability to climate change and enhanced adaptive capacity. Kavi Kumar (2009) contends the need to have region specific adaptation measures in farming as the key to address challenges of climate change. The Western Ghat region in southern India is one of the hotspots of climate change. The region witnessed two consecutive flood years in 2018 and 2019 resulting in intense rainfalls, destroying crops and livelihood settlements. Various studies report widespread loss of crops (IISR, 2019) and agro biodiversity (MSSRF, 2020) due to such climate change induced natural calamities in the region. Besides, studies also points to shift in the cropping patterns due to climate change in the high ranges (Merga and Alemayehu, 2019). However, studies are scarce with regard to the adoption of various adaption practices of farming communities. This study is an attempt to fill this gap. This study is framed with an objective to assess the knowledge and extent of adoption of climate resilient farming practices by the growers.

2. Methodology

The study was conducted in Wayanad, the hilly district in Kerala state of India. Wayanad is an agrarian district, lying in the high range zone of the Kerala.

Importantly, the whole district falls under the Western Ghat region, dominated by deciduous and evergreen forests and also recognized as a biodiversity hotspot (Kumar et al., 2011). Agriculture and allied sector contribute 43 per cent of Gross Domestic Product of the district and more than half of its population is directly engaged in farming to earn their livelihood. The gross cropped area is 78.59 per cent of the geographical area, mainly dominated by the cash crops (Department of Economics and Statistics, Government of Kerala, 2020). Further, 40 percent of the land area is under forest and tribal communities constitute 17 percent of the population of Wayanad (IIMK, 2006). The Kerala State Action Plan on Climate Change (2022-23) has identified Wayanad as one of the four climate change hotspots in Kerala, with a high degree of vulnerability. This is also confirmed by many of the previous studies (Rao et al., 2011, Kandiannan et al., 2014) particularly in the agricultural sector (Kumar et al., 2017; Radhakrishnan and Gupta, 2017).

The Agro Ecological Units (AEU) present in the region was considered for selection of the sampling villages as indicated below (Table 1). AEUs are the land units classified based on geomorphology, land use and soil variability. Of the 23 AEUs in the state, three are present in Wayanad viz. Northern High Hills, Wayanad Central Plateau and Wayanad Eastern Plateau (Kerala Agricultural University, 2020). This ensured inclusion of farming categories from all the AEUs in the study whose farming profiles are different. From each of these three AEUs 2 villages each were selected. Further, stratified random sampling technique was followed for the selection of households from these villages based on total number of farm households in these villages.

Table 1: Details of selected villages, number of respondents and AEUs

Sl. No	Name of the Village	Number of respondents	Agro Ecological Unit	Mean Annual temperature and Rainfall
1	Meppadi	20	Northern High Hills	26.2°C 3460 mm
2	Pozhuthana	49		
3	Meenangadi	44	Wayanad Central Plateau	22.6°C 2659 mm
4	Nenmeni	56		
5	Mullankolly	45	Wayanad Eastern Plateau	22.6°C 2658.9 mm
6	Pulppally	38		
	Total	251		

Various adaptation practices were listed out by reviewing the literature and by consulting relevant experts and field practitioners. The checklist was used while interviewing the respondents to know the listed adaptation strategies and its status of adoption. Score of 1 is given for adoption of a specific practice and 0 for no adoption. Adoption index is operationalized as the total score obtained by the respondents out of the possible score with regard to adoption of climate adaptation practices. The relationship between farm specific parameters and adoptions status was carried out to analyze their correlation.

3. Results and Discussion

Socio-economic Characteristics

Majority of the respondents were male farmers (92.03%) with agriculture as the main stay. The mean age is 58.37, with an annual income of 1.38 lakh rupees and land holding of 3.07 ha. Major crops prevalent in the region included plantation crops (Coffee, Arecanut, Banana), spices (pepper, ginger, turmeric) and field crops (Rice, vegetables). Besides, dairying/poultry was a subsidiary occupation for more than half of the respondents. The level of adoption of various adaptation practices by the respondents was investigated by using the checklist of such practices prepared and the results are presented below (Table 2).

Table 2: Practice of climate change adaptation practices by the respondents and constraints in adoption

Sl. No	Climate Change Adaption Practice	Practice of the Specific Adaptation Strategy		
		Yes	No	Main Constraints for Adoption
I	**Water Management practices**			
a	Use of drip irrigation systems	21	230	Lack of technical knowledge and cost
b	Irrigation scheduling of crops	130	121	Lack of technical knowledge
c	Water harvesting practices/ technologies	136	115	Limited subsidy provision
d	Construction of check dams and its management	91	160	Lack of technical knowledge and technical support
e	Water recycling techniques	177	74	-
II	**Land and Soil management practices**			
a	Minimum Tillage	248	3	-
b	Crop mulching	204	47	-
c	Contour bunding	209	42	Lack of technical knowledge and technical support
d	Fallow land cultivation	182	69	-
e	Farm diversification	176	75	Lack of incentives
f	Manure management practices (Biogas/ composting)	74	177	Lack of technical knowledge, cost and technical support
g	Raised bed system of planting	127	124	-
h	Grass stripping	56	195	-
i	Use of soil ameliorants	221	30	Lack of timely availability of inputs
j	Use of bio fertilizers and bio pesticides	174	77	Lack of timely availability of inputs, Poor access to input points

Sl. No	Climate Change Adaption Practice	Practice of the Specific Adaptation Strategy		
		Yes	No	Main Constraints for Adoption
k	Soil test based fertilizer application	13	238	Lack of timely availability of test results; Poor? technical knowledge
III	**Crop Management practices**			
a	Changing planting/harvest time	210	41	Lack of updated information
b	Cultivation of short duration varieties	21	230	Lack of technical knowledge; availability of varieties
c	Use of resilient local varieties	43	208	-
d	Crop diversification	250	1	-
e	Crop rotation	93	158	-
f	Cover cropping	1	250	Lack of seeds for cover cropping
g	Use of tolerant crop varieties	8	243	Lack of technical knowledge
h	Management of pest and diseases	132	119	Lack of technical knowledge support
i	Intercropping/ Mixed cropping	244	7	-
j	Practice of green manuring	224	27	-
k	Agroforestry	248	3	-
l	Split doses of fertilizer application	149	102	Lack of technical knowledge support
m	Hydroponics/ Aquaponics	1	250	Lack of technical knowledge, cost and technical support
n	Optimizing crop calendars	207	44	Lack of technical knowledge support
o	Low cost poly houses	11	240	Lack of technical knowledge, cost and technical support
IV	**ICTs and farming risk management practices**			
a	Weather based crop advisory	87	164	Generalized information, credibility of information
b	Subscription of crop insurance	63	188	Lack of technical formalities, cost, insufficient knowledge of its working mechanism
c	Use of mobile applications to access crop management information	4	247	Lack of technical knowledge
d	Use of early warning information for scheduling farm operations	79	172	Lack of technical support system
e	Knowledge about local adaptation strategies	28	223	-

From the results, it is evident that among the various adaptation strategies, higher proportion of farmers follows water, land and soil and crop management practices compared to ICT based and technological information. Further, majority (>80 %) of the farmers practice land and soil management such as minimum tillage, mulching, use of soil ameliorants as well as crop management practices like crop diversification, mixed cropping and agro forestry. Lowest adoption was observed with respect to soil test based fertilizer application, use of cover crops, tolerant varieties, use of new farm techniques like aquaponics/ hydroponics, low cost polyhouses and use of mobile applications for crop management. However, the constraints to adopt these adaptation practices differ from the findings of Rama Rao et al. (2018). According to them the major barriers were lack of weather information, labor scarcity, non-remunerative prices and non- availability of inputs. According to Jha and Gupta (2021), institutional factors such as extension services and technical knowledge support are key factors, influencing the likelihood of adopting climate smart farming practices. Some of the practices such as low level of subscription of crop insurance were identified as the potential drivers of climate change vulnerability of Kerala by other studies (DST, 2020).

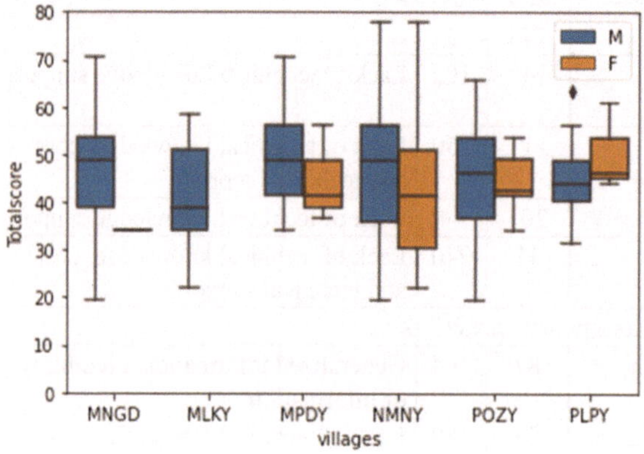

Fig. 1: Gender wise adaptation index score in the study villages

This box plot (Figure 1) shows the variation in the adoption index score of the respondents in various study villages. In majority of the study areas, male farmers tend to have higher median scores than the female farmers. The results are in line with the findings of Ebenehi et al. (2018).

Table 3: Status of specific adaptation mechanisms disaggregated by land holding status

Respondent Categories	Soil Testing	Training Participation	ICT use*	Membership in Farmer Organization	Livestock Rearing
Marginal farmers (Less than 1 ha) Count = 30	15 (50.00)	16 (53.33)	20 (66.67)	22 (73.33)	17 (56.67)
Small (1 to 2 ha) Count= 50	29 (58.00)	28 (56.00)	43 (86.00)	36 (72.00)	35 (70.00)
Large (>2 ha) Count= 171	102 (59.65)	116 (67.84)	148 (86.55)	120 (70.18)	123 (71.93)

*ICT Adoption: Any 3 or more ICTs

#Figures in the parenthesis indicate percentage

Further the study looked into adoption of specific adaptation strategies followed by the farmers, disaggregated by their land holding status (Table 3). From the analysis it is evident that more than two third of the respondents from all the categories use any 3 or more ICTs, have membership in farmer organizations and practice livestock rearing. Soil testing and training on climate resilient farming practices are followed by more than half of the participants. Besides, it is clear that membership in farmer organizations is a mechanism which is practiced by farmers of all categories to the same extent. To continue, difference between farmer belonging to small and large categories are less compared to marginal group, with regard to practice of specific adaptation measures. Finally the study investigated the correlation between selected farm specific parameters and adoption status of the respondents (Figure 2). Already studies have tried to explore such a relationship. For instance, Funk et al. (2020) reported that, the household head's education and gender, as well as the farm's size and the household's size, assets, livestock ownership, poverty status and use of extension services, are significantly correlated with the households' choices regarding adaptations to cope with climate change.

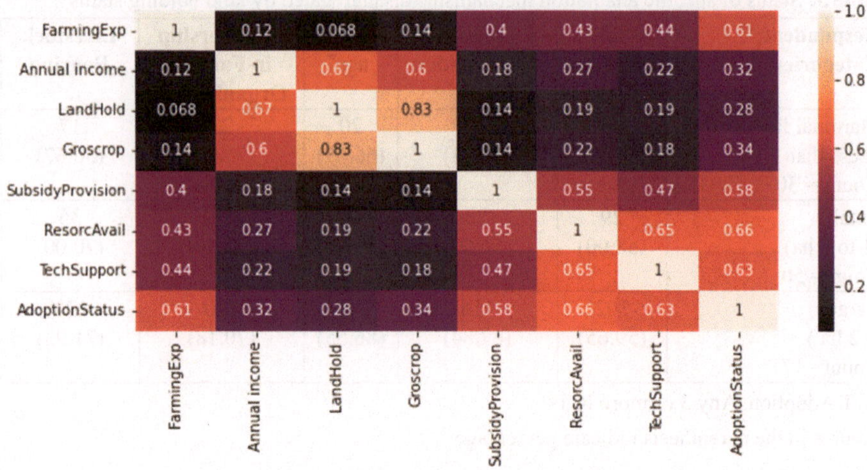

Fig. 2: Correlation matrix of selected variables and adoption index
**Note*: Adoption index is referred as the adoption status in the correlation matrix

It can be found that adoption of adaption practices was correlated with farming experience, availability of subsidy for the technology, ready availability of resources within the farm and technical support for adoption of the practice. The findings are mostly in line with the results of Senapati (2019). Further, in adopting climate smart practices like improved irrigation, better soil and crop management and mechanization, farmers prefer more governmental support as evident from the findings of Rama Rao et al. (2018).

Conclusion

Climate change is an imminent threat to socio-economic development globally and in India. Its impacts irreversibly affect rural population, and in the Western Ghat region, farmers try to cope up by adopting various management practices. From this study, it is evident that there is a variation with respect to the level of adoption of adaptation measures by the farmers. Higher proportion of farmers practice land, soil and water management practices whereas a less proportion tends to do so in case of ICT based and technological practices. Also the barriers to adopt climate adaptation strategies also differ depending on the type of the practice. Lack of technical knowledge support, timely availability of inputs, cost of the technology are some of the important concerns in this regard. Further gender difference was found to be prominent with regard to adoption of climate smart practices with male farmers bettering than their female counterparts. Factors like local resource availability, subsidy provision and technical support were found to be significantly correlated with the adoption of adaptation practices apart from the farming experience of the cultivator. Hence policy interventions should focus on strengthening institutional support in these areas. To improve the adaptive capacity of the society, policy level suggestions viz., schemes for crop and livestock diversification, soil and water conservation measures and integrated farming may be desired. In addition to that, capacity building programmes on natural disaster management and better adaptation measures may be given periodically to create awareness and share knowledge with the farming communities.

References

Akinnagbe, O. M., &Irohibe, I. J. (2014). Agricultural adaptation strategies to climate change impacts in Africa: a review. Bangladesh Journal of Agricultural Research, 39(3), 407-418.

Becerra, T. A., Middendorf, G., Campbell, A., & Tomlinson, P. (2016). Climate Change Challenges for Extension Educators: Technical Capacity and Cultural Attitudes. Journal of Extension, 54(6), Article 23. https://tigerprints.clemson.edu/joe/vol54/iss6/23

Department of Economics and Statistics, Government of Kerala, (2020). Data& Insights. https://www.ecostat.kerala.gov.in/

Department of Science and Technology (DST). (2020). Climate Vulnerability Assessment for Adaptation Planning in India Using a Common Framework. New Delhi: Department of Science and Technology, Government of India.

Deressa, T.T., Hassan, R.M., Ringler, C., Alemu, T., Yesuf, M., (2009). Determinants of farmers' choice of adaptation methods to climate change in the Nile Basin of Ethiopia. Global Environ. Change 19, 248–255.

Ebenehi, O., Ahmed, T. A., & Barnabas, T. M. (2018). Evaluation of extension services delivery for climate change adaptation by crop farmers in Niger State, Nigeria. Asian Journal of Agricultural Extension, Economics & Sociology, 27(1), 1-13.

Funk, C., Sathyan, A. R., Winker, P., & Breuer, L. (2020). Changing climate-Changing livelihood: smallholder's perceptions and adaption strategies. Journal of environmental management, 259, 109702.

Harvey, C. A., Saborio-Rodríguez, M., Martinez-Rodríguez, M. R., Viguera, B., Chain-Guadarrama, A., Vignola, R., & Alpizar, F. (2018). Climate change impacts and adaptation among smallholder farmers in Central America. Agriculture & Food Security, 7(1), 57.

Indian Institute of Management Kozhikode (IIMK). (2006). Wayanad Initiative. A situational study and feasibility report for the comprehensive development of Adivasi communities of Wayanad (Submitted to Scheduled Castes and Scheduled Tribe Development Department, Government of Kerala.

Indian Institute of Spices Research (IISR). (2019). Report on impact of rain induced natural calamity on spice crops in Kerala. http://spices.res.in/pages/report-impact-rain-induced-natural-calamity-spice-crops-kerala

James, R., Otto, F., Parker, H., Boyd, E., Cornforth, R., Mitchell, D., & Allen, M. (2014). Characterizing loss and damage from climate change. Nature Climate Change, 4(11), 938-939.

Jha, C. K., & Gupta, V. (2021). Do better agricultural extension and climate information sources enhance adaptive capacity? A micro-level assessment of farm households in rural India. Ecofeminism and Climate Change, 2(2), 83-102.

Kandiannan, K., Krishnamurthy, K. S., Gowda, S. J. A., and Anandaraja, M. (2014). Climate change and black pepper production. Indian J. Arecanut Spices Med. Plants. 16(4): 31-37.

Kavi Kumar KS. (2009). Climate sensitivity of Indian agriculture: do spatial effects matter? SANDEE working paper, Kathmandu; p. 45–09.

Kerala Agricultural University. (2020). Agoecologica Zones of Kerala. Delineation and Cropping pattern. Report submitted to Government of Kerala. Directorate of Research. Kerala Agricultural University.

Kumar, A. P., Lazarus, T. P., Santha, A. M., Brijit, J., & Manju, R. V. (2017). Impact of Climate Change on Black Pepper Production in Idukki and Wayanad Districts of Kerala. Int. J. Curr. Res, 9(6), 52960-52963.

Kumar NS, Aggarwal PK, Swaroopa R, Jain S, Saxena R, Chauhan N. (2011). Impact of climate change on crop productivity in Western Ghats, coastal and northeastern regions of India. Current Science. 101(3), 332-341.

Merga, W and D Alemayehu. (2019): "Effects of Climate Change on Global Arabica Coffee (Coffea arabica L) Production", Greener Journal of Plant Breeding and Crop Science, Vol. 7, No. 1, pp. 23-30.

MSSRF. (2020). Assessment of Agro biodiversity loss due to flood in Wayanad. Conservation strategies for resilient varieties. Retrieved from https://www.keralabiodiversity.org/images/2019/November/Flood_Report/Final_Report_MSSRF.pdf

Mwazi, F., &Ndokosh, J. (2011). Effective communication of climate change by Extension agents. Agricola. Pp. 20-23.

Radhakrishnan, A., & Gupta, J. (2017). Vulnerability of dairy-based livelihoods to climate variability and change: a study of Western Ghats region in Wayanad, Kerala. Current Science, 113(1), 123.

Rama Rao, C. A., Raju, B. M. K., Rao, A.V. M. S., Rao, K. V., Ramachandran, K., Nagasree, K., Samuel, J., Ravi Shankar, K., Srinivasa Rao, M., Maheswari, M., Nagarjuna Kumar, R., Sudhakara Reddy, P., Yella Reddy, D., Rajeshwar, M., Hegde,S., Swapna, N., Prabhakar, M. and Sammi Reddy, K. (2018). Climate Change Impacts, Adaptation and Policy Preferences: A Snapshot of Farmers' Perceptions in India. Policy Paper 01/2018. ICAR-Central Research Institute for Dryland Agriculture, Hyderabad, India. 34 P.

Rao, G. S. L. H. V. P., Rao, A. V. R. K., Krishnakumar, K. N., and Gopakumar, C. S. (2011). Climate change projections and impacts on plantations in the humid tropics. In: Rao, G. S. L. H. V. P. (ed.), Climate Change Adaptation Strategies in Agriculture and Allied Sectors (1st Ed.). Scientific Publishers, India, 330p.

Senapati, A. K. (2019). Assessing the vulnerability of agricultural households to covariate and idiosyncratic shocks: a case study in Odisha, India. Climate and Development, 1-15.

Zamasiya, B., Nyikahadzoi, K., &Mukamuri, B. B. (2017). Factors influencing smallholder farmers' behavioural intention towards adaptation to climate change in transitional climatic zones: A case study of Hwedza District in Zimbabwe. Journal of environmental management, 198, 233-239.

5

Climate Change and Agriculture Effects and Adaptive Responses in the Western Ghats

Archana Bhatt, Shakeela V., Habeeb Aand Divya P.R.

M S Swaminathan Research Foundation – Community Agrobiodiversity Centre, Wayanad, Kerala

Abstract

Asia Pacific region is one of the most climate change vulnerable regions in the world and India is no exception. Agriculture being highly dependent on climatic condition is like gambling at the hands of weather disasters. In this context, Adaptation at local level is a critical issue as local actors are the ones who realize the severity of climate change. Hence, a study titled ": Effects and Adaptive responses in the Western Ghats" was conducted with an objective to focus on climate change adaptation research. The study documented the effects of climate change on agriculture and the adaptation& mitigationefforts.Quantitative and qualitative approaches were usedto elicit information from a total of 270 farmers across 3 Agro Ecological Units (AEUs) in Wayanad, Kerala. Findings revealed that majority (92.96%)of the farmers were aware about climate change and observed shift in weather patterns (98.14%). Study reported that over a span of past decade, farmers observed a range of effects in relation to climate variability especially increase in average temperature (95.92%), prolonged dry spells (76.29%), fluctuation in onset of monsoon (94.81%), uneven rainfall distribution (94.07%), flood/drought incidences (64.81%), decline in ground water (65.18%), shrinking water bodies (63.33%) and newer pest resurgence (64.81%). Across the three AEUs, Southern High Hills reported higher percentage in terms of climate variability perceived by the farming community. Farmers followed various adaptation and mitigation measures in response primarily change in sowing/planting dates, crop diversification, crop rotation, mixed cropping/ inter cropping, incorporation of crop residues, mulching, application of bio-fertilizers, raised beds and planting of native trees. It was reflected through the study that majority of farmers followed adaptation & mitigation practice unintentionally rather than knowingly the benefits it can provide in the context of climate variability A thorough understanding of the current status of adaptation mechanisms at farm level will guide for fruitful implementation of adaptation measures in future.

Keywords: Climate Change, Adaptation, Mitigation, Agriculture, Western Ghats

Introduction

We are faced with the dual challenges of Climate change and food insecurity irrespective of the countries and regions across the globe. Asia Pacific region is one of the most climate change vulnerable regions in the world (Prabhakar & Matsumoto, 2010) on account of its relatively higher & dense population depending on climate related sectors and living in climate change vulnerable locations. In particular, the agriculture production system is facing serious repercussions due to increasing weather vagaries since past few decades. Agriculture being highly dependent on climatic condition is like gambling at the hands of erratic rainfall or droughts leading to crop loss on a regular basis in terms of reduced yield, increased pest resurgence and lack of safety net in case of weather based disasters all of which prominently affects the small holder farm families. Indian subcontinent has been facing extreme climatic events since the past decades in form of floods, erratic rainfall, high temperature, shrinking water sources, etc. which have caused deprivation in agriculture in economic and non-economic terms. In a series of researches and studies conducted over the years; the need for enhanced adaptation research and policymaking capacity in developing Asia has been reflected. Hence, adaptation research has to take an important place at global, national and local level.

Adaptation at local level is the most critical issue as local actors are the ones that realize the severity of climate change (UNFCCC, 2009) i.e. the farming community. Studies conducted across India demands the need for tailor-made climate adaptation policies and strategies in relation to the local socioeconomic conditions, geographical setting, and weather based challenges faced by the farmers (Varadan and Kumar, 2015; Harvey et al., 2018). Therefore, there is a need for taking up adaptation research at a more local or regional level to develop efficient adaptation strategies.

Besides, vulnerable communities do take up certain adaptation mechanisms to cope with the effects of climate change and these mechanisms may differ both spatially and temporally that need to be studied. Hence, a thorough understanding of the adaptation responses and the effect of climate change on farming needs to be taken up at micro level.

In line with the above mentioned points, a study was conducted in Wayanad, Kerala to study the adaption & mitigation response of farmers towards the climate variability effects on various crops. Information collection was also validated through secondary data available and interview with experts.

Methodology

Wayanad lies in the high range zone of Kerala and is a rich agro-biodiversity hot spot nestled in the Western Ghats. Various studies have reported the proneness of the region to climate change (Kandiannan*et al.,* 2014). Flood induced from heavy rainfall in 2018 and 2019 incurred significant losses especially to the farming community of Wayanad. Increasing incidences of climate variability over the past few years demands the need of a climate change oriented study in the district. Besides, the district is identified as one of the Aspirational districts by NITI *Aayog* for special developmental attention in India (Aspirational District Programme-Wayanad, 2018-22) which demands the acute need for research & development in the district. Besides, Wayanad district is one of the highly vulnerable districts of Kerala towards climate change (KSAPCC, 2022).

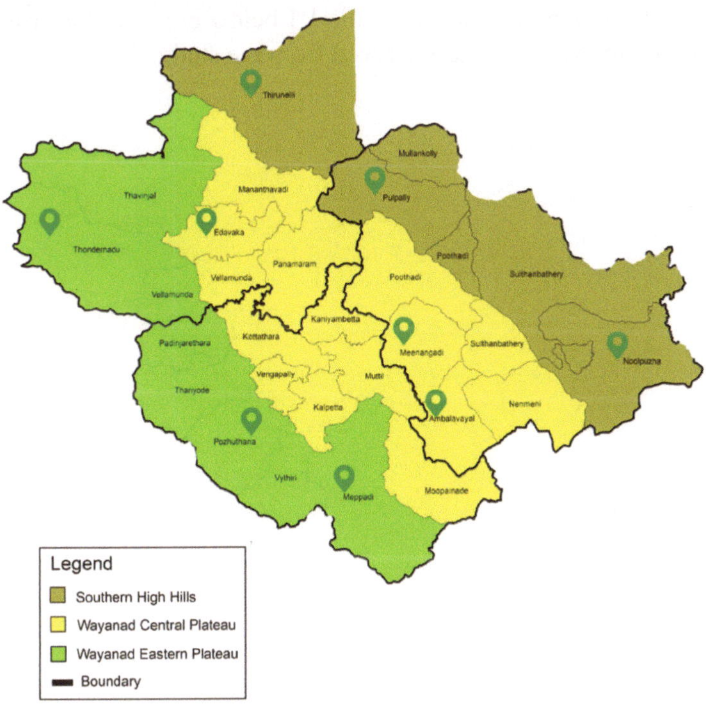

Fig. 1: Map of Wayanad district indicating study area

Three Panchayaths from each Agro-ecological unit were randomly selected for the study (Figure 1). 30 farmers from each Panchayath were randomly selected i.e. a total of 270 farmers were surveyed for the study. Semi Structured Interview Schedule, Focused Group Discussion and Key Informant Interviews were conducted for the data collection. Appropriate statistical tools and techniques were used for data analysis and interpretation.

Result & Discussion

Agriculture System in Wayanad

Wayanad is a hilly district of Kerala nestled amidst the Western Ghats with a continued history of agriculture. Various communities have been engaged in cultivation in the district from paddy cultivation to plantation and homestead farming. Tribal communities in the district practiced shifting cultivation which later transformed to settled agriculture through paddy cultivation and plantation crops. With time, tea, coffee, pepper became prominent crops with other important crops like banana, ginger and other spice crops and vegetables. Agriculture system in Wayanad can be broadly defined as paddy based system (paddy +fallow/cowea), paddy and banana system, coffee based agroforestry and agro-biodiversity rich homesteads. Table 1.1 below indicates the crop and livestock diversity in the district as observed from the study

Table 1: Agriculture in Wayanad district

Particulars	Cereals, Millets & Pulses	Vegetables	Fruits	Plantation & Spice crops	Roots & Tubers
Major crops	Paddy, Cowpea	Tomato, Chili, Brinjal, Cabbage, Cauliflower, Okra, Cucurbits, Leafy greens, Carrot, Beetroot, Chillies, Drumstick	Banana, Sapota, Mango, Guava, Avocado, Gooseberry, Aonla, Custard apple, , Jackfruit, Passion fruit, Papaya, Rose apple, Litchi, Lemon,	Coffee, Rubber, Arecanut, Coconut, Cardamom, Pepper, Arecanut, Curry leaves, Tea	Turmeric, Ginger, Colacasia, Tapioca, Dioscorea, Alocasia,
Minor crops	Ragi, French bean, Winged bean, Amara payar	Broccoli, Chinese cabbage, Chinese potato, Breadfruit	Mangosteen, Orange, Garcinia, Pomegranate, Egg fruit, Rambutan, Star fruit, Jamun, Strawberry, Tamarind	Vanilla, Cashew nut, Clove, Cocoa, Nutmeg	Sweet potato, Arrowroot
Livestock & Fisheries	Cattle (Traditional & Cross bred), Buffalo, Goat, Pig, Poultry (Hen, Turkey, Quail, Duck), Rabbit, Fisheries (Catla, Tilapia, Rohu)				

Nevertheless, Wayanad being a biodiversity hot-spot; is home to range of native diversity of various crops including pepper, banana, tubers, rice and more. This diversity has the potential to play a key role in the context of climate change since a range of diverse crops have climate resilient characteristics. For example, tubers can thrive in a water scarce condition, traditional rice varieties are climate resilient like Veliyan (drought & flood tolerant), Kalladiaran (drought tolerant & short duration) to name a few (Gopi & Manjula, 2018). Besides, the region harbours various wild edible and wild crop relatives that can be utilized for breeding and improvement in future.

Climate Variability in Wayanad

Wayanad has been facing climate variability related challenges since decades which has also been reflected through various studies and reports. Throughout the study, farmers reported various climate-induced threats and challenges mainly uneven rainfall pattern especially erratic rainfall, prolonged dry spell, fluctuation in the onset of monsoon and fluctuation in the average temperature and rainfall. It was also observed through the survey that Noolpuzha region from the Southern High Hills was experiencing from drought effect in the summers with prolonged average dry spell. Majority of the surveyed farmers asserted a gradual rise in warm temperature, decline in overall rainy days but more erratic & uneven patterns (Figure 2 and Figure 3) which also lines with the reported weather data in the district. Majority (92.96%)of the farmers were aware about climate change and observed shift in weather patterns (98.14%).

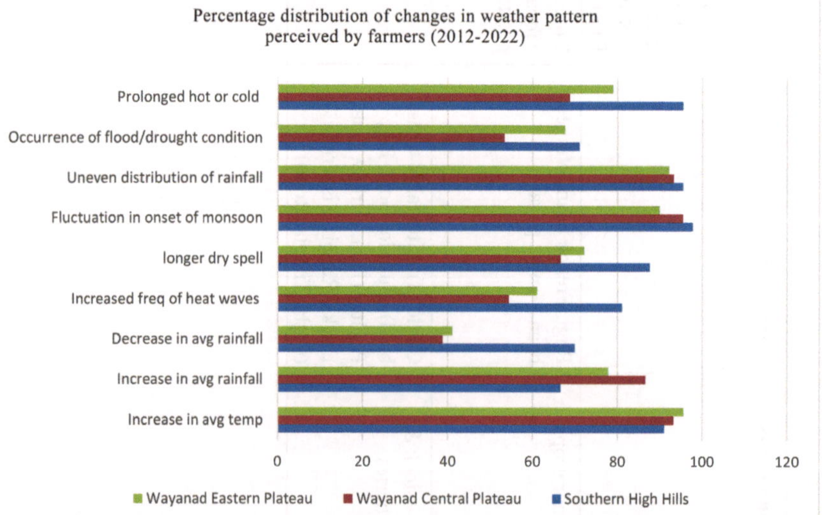

Fig. 2: Percentage distribution of changes in weather pattern as perceived by farmers

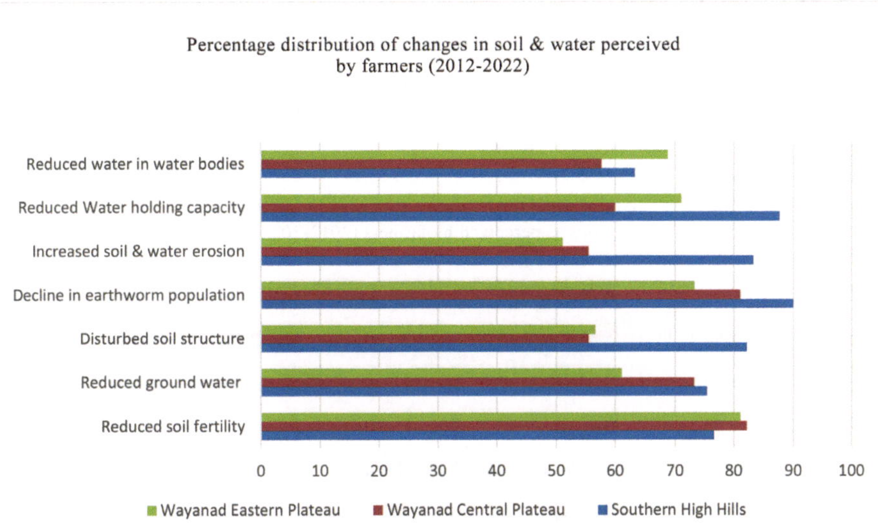

Fig. 3: Percentage distribution of changes in soil & water as perceived by farmers

Effects of climate variability on crops

Climate change refers to the "variation in either mean state of the climate or in its variables persisting for a long and an extended period (decades or longer). While climate variability refers to sudden and discontinuous seasonal or monthly or periodic changes in climate or its components without showing any specific trend oftemporal change" (IPCC 2013; Venkatramanan and Shah 2019). Wayanad district has been dwindling with climate change as evident from the meteorological data over the years reported through various studies (Kumar, 2017). When it comes to climate variability, over the past few years, there have been observable changes in the climate especially rainfall pattern and temperature in the district as reflected from discussion with farmers. These changes in turn are affecting the farming system to a significant degree as perceived by the farming community.

It is also worth noting that over the past decades, human wildlife conflict has been on the rise which might be on account of fewer resources in the forest surrounded Panchyaths like Noolpuzha region. Drying up of natural water and other resources might have led the wild animals like elephants, wild boar and other to frequent the farmers' fields as reflected in the discussion with farmers. This in turn more than often leads to crop damage and destruction in forest surrounded farm families in the district.

Figure 4 and Table 1.2 showcases the various effects of climate change as perceived by the farming community in Wayanad district. As observed through the survey and discussions with farmers, rice, pepper and coffee are the most affected crops from climate variability when it comes to cash crops.

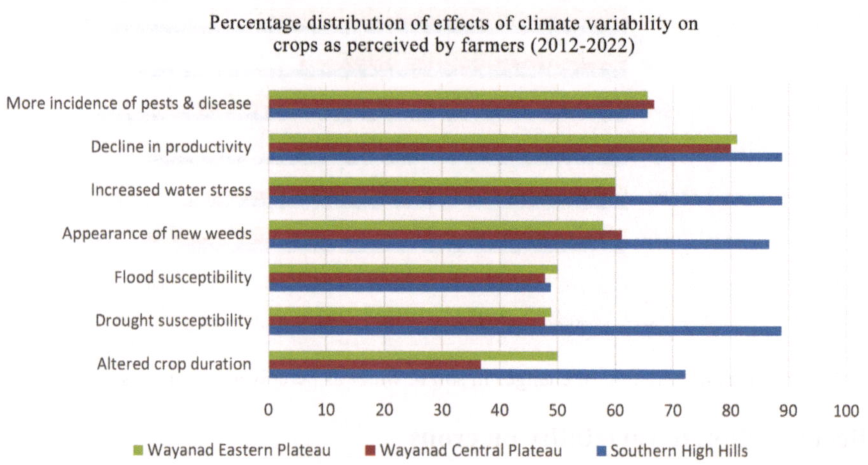

Fig. 4: Percentage distribution of effects of climate variability on crops as perceived by farmers

Rice: Majority of the farmers reported that occurrence of blast disease became more prominent after the floods in 2018 and 2019 in the region. There was significant loss to paddy due to the floods but farmers adapted through the cultivation of High Yielding Varieties to recover the loss. Traditional varieties namely '*Kalladiaran*' and '*Veliyan*' were especially considered resilient to weather vagaries during the time. *Veliyan*was observed to be flood tolerant while *Kalladiaran* ensured yield after flood related losses due to its short duration. Unseasonal and erratic rainfall has also led to increased incidence of Brown Plant Hopper in the crop. New pests like blue beetles (*Leptispapygmaea*) were also observed in the paddy field over a span of last 10 years. Though, rice cultivation has declined in summers, higher temperature is affecting the existing cultivation of *Punja* season rice cultivation due to water scarcity and continued reduced ground water resource.

***Pepper*:** Quick wilt and slow wilt is one of the major issues in pepper cultivation which is also affected by the climate change. Besides nematode being a primary cause, quick wilt and slow wilt has been progressing in the district with climate variability being one of the major cause. Uneven rainfall pattern coupled with water scarcity in certain regions is leading to wilt, berry drop and reduced berry size as perceived by the farmers. A study conducted by Kumar et al (2017) on impact of climate change on pepper production indicated that increased temperature resulted decrease in pepper production in the district. Moreover, erratic rainfall coupled with high Relative humidity creates congenial environment for disease and pest incidences in the crop causing lower production.

***Coffee*:** Flowering in coffee is highly correlated with the rain showers hence any unevenness in the rainfall pattern affects the flowering and fruiting in the crop. Summer showers are essential for flowering which need to be backed by rain showers for fruit or berry development. Any climate variability in this context will affect the coffee production.

***Ginger and Banana*:** In case of crops like ginger and banana, flooding or water logging can have severe repercussions on the rooting system. Erratic and heavy rainfall occasions lead to high incidence of root rot and other disease occurrence.

***Other crops & vegetables*:** Major effects on vegetable crops experienced by the community was found to be wilting due to increase in temperature. Majority of the farmers reported that high day temperature has been affecting the crops due to increased water stress.

Table 2: Effect of climate change on crops: Farmers' viewpoints (Last 10 years)

Crop	**Southern high hills (Tirunelli, Pulpally, Noolpuzha)**	**Wayanad central plateau (Ambalavayal, Meenangadi, Edavaga)**	**Wayanad easternplateau (Thondernadu, Meppadi, Pozhuthana)**	**Management options followed by the farmers**
Paddy	Bugs and BPH attack Stem Borer and army worm Wilting in summer (Pulpally) Blast after flood Stem borer and chaffy seeds Appearance of new pests (Blue beetles)	Stem borer and bugs attack Blast Yellowing and blight Leaf folder Appearance of new pests (Blue beetles)	Caterpillar attack especially stem borer Yellowing of leaves Bugs Blast Increased weeds	Paraka (*Ficus hispida*) leaves for removing worms Tobacco decoction Fish amino solution Organic manure Ash and fertilizers Biofertilizers Ant powder Irrigation management Weedicides
Pepper	Quick wilt in monsoon Root rot Berry drop Decreased berry size Decline in productivity	Quick wilt in monsoon Root rot Berry drop Yellowing in summer Plant burn	Quick wilt in monsoon Delayed spike formation Scale insect Berry drop Root rot & Yellowing after monsoon	Cow dung application, Neem cake, Replanting, Bordeaux mix, Trichoderma Bordeaux mixture Tobacco formulation Lime application Pseudomonas, Neem oil spray Micro nutrient Recommended dose of fertilizers
Coffee	Change in flowering time Berry drop Stem & berry borer Wilting and yellowing	Change in flowering time Berry drop Stem borer Root rot Wilting	Change in flowering time Berry drop Black rot Leaf fall Wilting Coffee borer Dying of whole plant	Pesticides Organic manure Water management Replanting, Biofertilizers Insect traps

Ginger	Mahali Root rot Borer	Mahali Root rot Borer	Borer Root rot Mahali Leaf curling Drying	Copper sulphate+Lime application Tobacco decoction Pesticides, Bordeaux mix Lime & Bleaching powder
Areca nut	Mahali and fruit drop Drying up of entire plant	Mahali Bud rot	Fruit drop Top most portion dead Summer Mahali Wilting	Organic manure Bordeaux mixture Pesticide application/Urea & Lime, replanting, Bordeaux mix
Coconut	Mandari (Eriophyd mite) Rhinoceros beetle	Mandari (Eriophyd mite) Rhinoceros beetle Top portion dead	Mandari (Eriophyd mite) Rhinoceros beetle Top portion dead	Sand+Salt mixture or Neem cake+Salt Salt+Copper sulphate mix Bordeaux mix and lime aplication
Banana	Wilting Weevil, borer and skipper Leaf curling	Bunchy top Weevil and borer Root rot n monsoon	Bunchy top Yellowing Panama wilt Wilt Caterpillar Root rot Rotting from inside and black colouration	Irrigation management Urea + Cow dung + water Tobacco decoction Pesticide application Humic acid application Bordeaux mixture
Vegetables	White fly Wilting in summer Mealy bugs and other bugs Leaf curling (Aphids+Ants)	White fly Wilting in summer Mealy bugs and other bugs Leaf curling (Aphids+Ants)	Worms and bugs Mealy bugs Common diseases Wilting White fly	Tobacco pesticide Cow dung Ash+Turmeric powder Water management Rice water+Ash+Cowdung Garlic+Chilli solution Soap water+Cow dung Trichoderma and Pseudomonas
Fruits	Wilting in Jackfruit	Fruit drop in avocado, wilting in Jackfruit	Wilting	Bio fertilizers and bio pesticides

Adaptive Responses to Climate Change

Throughout the study, farmers reported that climate variability has had a significant impact on agriculture and affected various cropping systems predominantly paddy, coffee, pepper and ginger as reflected above. To cope with the effects of climate variability, farmers in the district followed various adaptation and mitigation practices including mulching, changing the sowing time, usage of bio-fertilizers and organic manure, crop and farm diversification, crop & variety rotation, short duration varieties, alternate livelihoods, and more. The following figure (Figure 5) highlights the major climate change adaptation practices followed by the farmers in Wayanad. Many farmers either adopted alternate livelihood options from farm to non-farm activities or were willing to practice the same in future to ensure secured livelihood.

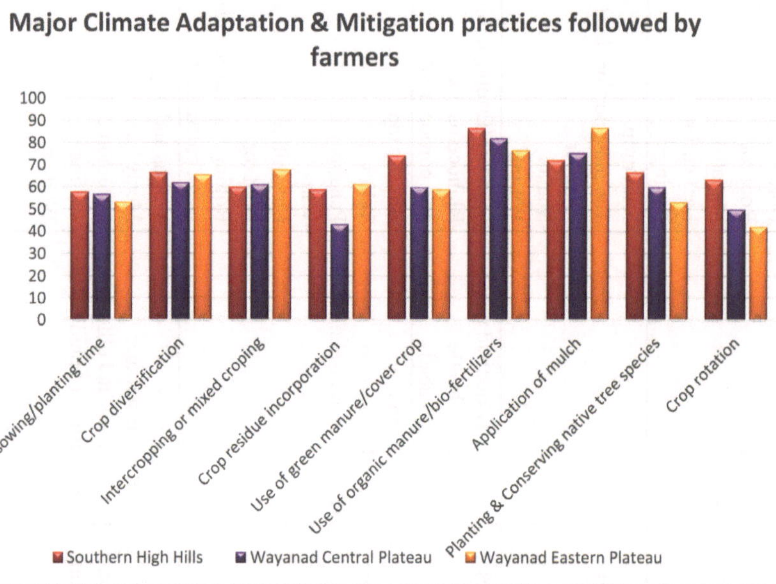

Fig. 5: Major adaptation & mitigation practices followed by farmers in Wayanad

***Changing the sowing or planting time*:** Due to increasing uneven pattern of rainfall and fluctuation in the onset of monsoon, majority of the farmers switched to changing the sowing or planting time especially in case of rice nursery preparation. Many farmers reduced the duration between nursery initiation and transplanting i.e. from the conventional 25-30 days to 20-22 days to cope with unprecedented rains.

***Crop diversification*:** More than adaptation to climate variability, most of the farmers in the district follow homestead farming system which allows them crop diversification ranging from paddy, coffee, native trees, banana & other plantation and vegetables in the homesteads.

***Crop residual incorporation*:** After the harvest of crops such as paddy, the residuals are incorporated into the soil via ploughing and helps enrich the soil.

***Green manure/cover cropping*:** Green manuring is practiced by the farmers in the district either by applying tree leaves such as *Glyricidia* sp. or by cultivating legume crops like cowpea after paddy harvest.

***Use of organic manure/bio-fertilizers*:** Even though many farmers do use inorganic fertilizers; organic fertilizers such as cowdung is an indispensable part of soil nutrition in all the crops. Besides, other form of bio-fertilizers and formulations such as Trichoderma and Pseudomonas are also used by majority of the farmers.

***Application of Mulch*:** Mulching is practiced by a majority of farmers to conserve moisture after planting any crop especially ginger, cowpea and others.

***Native tree species*:** Many traditional farmers as part of their cultural heritage plant and conserve the native tree species in their homesteads. Old aged trees form habitat for the biodiversity and also help sequester carbon in long term.

***Crop rotation*:** Crop rotation is also one of the primary adaptation practice followed by the farmers in various seasons and crops e.g. Paddy–cowpea, Paddy-banana, etc. Besides crop rotation, many farmers follow variety rotation as well especially in case of paddy which is also coupled with the change of field site annually or biannually to manage pest & disease occurrence.

The study conducted across the three Agro ecological units of Wayanad district shows that majority of farmers adapted to climate change or followed adaptation & mitigation practice unintentionally rather than knowingly the benefits it can provide in the context of climate variability. Besides the mentioned adaptation & mitigation practices, few farmers did follow other options such as improved rain-water harvesting practices, renovating ponds in the forests, constructing soil and water conservation structures like countours, bunds and trenches, switching to high yield varieties, crop insurance, increased application of organic fertilizers and pesticides, farm diversification and more. The farming community inculcated several changes in their farming system which were also covered under climate adaptation or mitigation practices. It is also important to note that some of the adaptation options might have negative consequences if not managed in a regulate manner especially higher dose of fertilizers & pesticides, use of ground water resources.

Conclusion

Climate variability has been a serious concern for the agriculture sector primarily in the recent years while significantly affecting the marginal and small holder farm families in developing and underdeveloped nations. India is also facing the brunt of uneven weather patterns over the past decades in form of floods, drought and erratic rainfall. These weather vagaries are responsible for a range of damage and destruction to crops and livestock or in its entirety the overall agro-ecosystem. In concern with the effects & impacts of these changes, it becomes imperative to take research and development actions on the effect of climate change especially at local level. This study specifically looked into the effects of climate variability on agriculture system as a whole inclusive of the effects on water, soil & crops based on farmers' perspective. Studies like this can throw light on the existing adaptation practices and can help in developing research & development action plans at local level. The study also reflected that many adaptation practices followed by the farmers were rather unintentional than being practiced based on their role in climate change adaptation. This also showcases the lack of awareness and knowledge on climate adaptation & mitigation practices among the farming community.

References

Gopi, G., & Manjula, M. (2018). Speciality rice biodiversity of Kerala: need for incentivising conservation in the era of changing climate. Current Science, 997-1006.

Harvey, C. A., Saborio-Rodríguez, M., Martinez-Rodríguez, M.R., Viguera, B., Chain-Guadarrama, A., Vignola, R., & Alpizar, F. (2018). Climate change impacts and adaptation among smallholder farmers in Central America. Agriculture & Food Security, 7(1), 1-20.

IPCC (2013) Summary for Policymakers. In: Stocker TF, Qin D, Plattner G-K, Tignor M, Allen SK, et al.., editors. Climate Change 2013: The Physical Science Basis. Contribution of Working Group I to the Fifth Assessment Report of the Intergovernmental Panel on Climate Change. Cambridge, United Kingdom and New York, NY, USA: Cambridge University Press

Kandiannan, K., Krishnamurthy, K. S., Ankegowda, S. J., & Anandaraj, M. (2014). Climate change and black pepper production.Indian Journal of Arecanut, Spices & Medicinal Plants, 15(6), 30-37.

Kerala State Action Plan on Climate Change. https://envt.kerala.gov.in/wp-content/uploads/2022/12/Kerala-State-Action-Plan-on-Climate-Change-2.0.pdf

Kumar, A. P., Lazarus, T. P., Santha, A. M., Brijit, J., & Manju, R. V. (2017). Impact of Climate Change on Black Pepper Production in Idukki and Wayanad Districts of Kerala. Int. J. Curr. Res, 9(06), 52960-52963.

Prabhakar, S. V. R. K., & Matsumoto, K. (2010). Mainstreaming Climate Change Adaptation in Agriculture and Water Sectors in the Asia-Pacific Region: Current Status, Issues, and Way Forward. Asian Journal of Environment and Disaster Management, 2(4).

UNFCC, U. N. (2009). Kyoto protocol reference manual on accounting of emissions and assigned amount. eSocialSciences.

Varadan, R. J., & Kumar, P. (2015). Mapping agricultural vulnerability of Tamil Nadu, India to climate change: a dynamic approach to take forward the vulnerability assessment methodology. Climatic change, 129(1), 159-181.

Venkatramanan, V., & Shah, S. (2019). Climate smart agriculture technologies for environmental management: the intersection of sustainability, resilience, wellbeing and development. In Sustainable green technologies for environmental management (pp. 29-51). Springer, Singapore.

6

Protected Cultivation Technologies Adoption and Constraints Perceived by Farmers

B.S. Meena and Ravi K. Meena

Agricultural Research Station, Sriganganagar
S.K. Rajasthan Agricultural University, Bikaner, Rajasthan

Abstract

The investigation on adoption of protected cultivation technologies- polyhouse and constraints perceived by farmer of Sriganganagar was undertaken in the year 2021-22 with a case study mode. A detailed survey was made of farmers having polyhouse structure for cultivated crops from the year 2005-06 to 2020-21 of Sriganganagar district. The major aspects on which information were collected includes extent of adoption of polyhouse-PCT, impact on socio economic status and constraints perceived. The investigation found that small holding farmer (less than 2 hac.) was most successful in polyhouse cultivation as compare to large farmer. Polyhouse owner mostly depends on fellow colleagues for technical knowhow of adoption crops. Study found that polyhouse farmer have not started cultivation of vegetables linked to certain factors of failure mainly due to lure to subsidy, for enhanced socio-economic status, lack of technical knowledge, lack of monitoring by horticulture department, Company monopoly- material quality, after care, insurance terms, design defaults, lack of technical backup, limited choices of vegetables- cucumber, Damage due to weather factors- air storms, diversification to other enterprises. Farmers who got sanction of polyhouse should ensure to undergone trainings on polyhouse cultivation aspects and management for success in this mode of cultivation.

Keywords: polyhouse, adoption, vegetable, constraints

India is endowed with diverse agro- climatic conditions which facilitate to cultivate various horticultural crops all through the year in one part of the country or others, but the quality of fruits under open field conditions particularly in high value crops (vegetables & flowers) by and large fall short of domestic and international markets standards. Some time there no guarantee of consistent production of from open cultivation, as the crop is exposed to number of environmental factors that changes frequently. Hence it is imperative to increase the productivity and quality of produce to meet the demand of quality conscious consumers. A break through in production technology that integrates market driven quality parameters with the production system simultaneously ensuring a vertical growth is warranted. In this context, most important technology is the Controlled Environment Agriculture (CEA) i.e. Protected Cultivated Technology (PCTs) such as greenhouse, net house, glasshouse and low tunnel. "Protected cultivation" means modification of plant environment to achieve optimum plant growth avoiding the harmful effect of climatic factors.

The study area Irrigated North-Western Plain (1b) zone covers a geographical area of 2.06 million hectares, comprising the districts of Sriganganagar and Hanumangarh of Rajasthan State. It is located between 28"4' to 38"6' North latitude and 72"3' to 75"3' East longitude. The states of Punjab and Haryana forms North-Eastern boundary, Churu and Bikaner districts of Rajasthan forms south boundary and international border of Pakistan forms the North and North-Western boundary of the zone. Soils of the zone are mostly light to medium textured, calcareous, well drained, deep having low nitrogen, low to medium phosphorus and medium to high potash.The zone has extreme climatic conditions with scorching summer, cold winter and mild rainy season. Dust storms during summer, frosty winter nights and ground fog are some of the typical features of weather hazards. The rainfall is relatively low in western parts as compared to eastern part of the zone. The rainfall in the zone ranged from 200 to 590 mm during the period from 1981 to 2009 with an average value of 325 mm. Most of the rainfall is received from July to September. The temperature in the zone fluctuates from as low as 10 C to as high as 490 C. January is the coldest and June is the hottest month of the year. The average monthly pan evaporation fluctuates from 54 mm in January to 335.4 mm in June, exceeding to 2000 mm annually.

The present investigation have been planned to address the above points with the following specific objectives

1. To study the profile of PCT farmers
2. To study the extent of adoption of protected cultivation technology among selected farmers

3. To study the impact of PCTs on socio-economic status of selected respondents
4. To identify the constraints and perspective for protected cultivation in the region.

Research Methodology

A research design is the overall plan or programmes to undertake the study. It includes an outlines of what the researcher will do from writing the hypothesis and its operational implications to the final analysis of data collected from primary and secondary sources. An ex-post facto research design shall be used for the present study as the manifestations of the variables presumably had already occurred and there was no scope for manipulation of any variables selected in the investigation. The protected cultivation structures namely polyhouse shall be taken under the investigation and proportionate farmers were selected randomly. Five farmers under each selected PCTs from the chosen district (Sriganganagar) have been selected as respondents to collect the primary information desired under the investigation. The information collected in a pre-designed interview schedule includes profile, PCT cultivation practices and constraints have been considered as bottle necks in the PCT were collected. The collected data will be subjected to statisticalanalysis and interpretation were use in preparing the research findings of the investigation.

Research Findings

Profile of the farmers: The selected respondents under the investigation were of middle age, having secondary to graduated level education, possessing large and very large size of holding and having training experience in precision farming. There was no major difference on profile of successful and failure farmers as far as polyhouse management is concerned. The major difference was degree of involvement of polyhouse owners in cultivation aspects.

Table 1: Brief particulars of Polyhouse Farmers (successful cases)

Sr. No.	Name of farmers	Year of polyhouse establishment	Size of polyhouse	Status
1.	Munsi Ram Chak KSD, Raisinghnagar	2019-20	2000 Sq.m	Functioning
2.	Deep Chand 10 A , SGNR	2018-19	2000 Sq.m.	Functioning
3.	Pramod Kumar 15 Z	2017-18	4000 Sq.m.	Functioning
4.	Vikram Bishnoi 1 GBD, Gharsana	2020-21	4000 Sq.m.	Functioning
5.	Prithvi raj 45 PS	2019-20	2000 Sq.m.	Functioning

Table 2: Brief particulars of Polyhouse Farmers (Unsuccessful case)

Sr. No.	Name of farmers	Year of polyhouse establishment	Size of polyhouse	Status	Remarks
1.	Anup Soni D/o Shyam Kumar 4 KM,Gharsana	2017-18	4000 Sq.m.	Structure exist	Not under cultivation
2.	Shyam kumar 4 KM, Gharsana	2018-19	4000 Sq.m.	Structure exist	Not under cultivation
3.	Ravi S/O Bheraram 26 PS, Raisinghnagar	2018-19	2000 Sq.m.	Structure exist	Not under cultivation
4.	Dinesh kumar 87 GB	2017-18	4000 Sq.m.	Structure exist	Not under cultivation
5.	Kulvinder Singh 15 PS Raisinghnagar	2016-17	4000 Sq.m.	Structure exist	Sublet for cultivation

Attributes of successful polyhouse farmers: Under the investigation visited respondents were asked on their responses on pre-identified success attributes and their visible impact on their overall performance of polyhouse management. The major contributing attributes were self- motivated, good technical knowledge, higher economic motivation, strive to enhance socio-economic status, hard work put up by self & his family, good liaisons with line departments, self managed, not lure for subsidy, received training on Polyhouse and its maintenance and possesses entrepreneurship attributes.

Identified constraints in polyhouse based cultivation: On the other side of the investigation, the responses from polyhouse owners did not manage well for which certain attributes might be responsible includeslure to subsidy, lack of technical knowledge, lack of monitoring by horticulture department, company monopoly- material quality, after care and insurance cover of polyhouse structure design defaults, lack of technical backup, limited choices of vegetables- cucumber, damage due to weather factors- air storms and diversification to other enterprises.

Conclusion

The investigation found that small farmer holding (less than 2 hac.) was most successful in polyhouse cultivation as compare to large farmer. It was found that polyhouse owner who himself involve in carrying out most of the cultivation practice and manage the polyhouse have received higher income and success in polyhouse based farming as compare to polyhouse owner who depends on other for cultivation operations and managerial aspects. Polyhouse owner mostly depends on fellow colleagues for technical knowhow of vegetable crops.Farmers who got sanction of polyhouse should ensure to undergone trainings on polyhouse cultivation aspects and management for success in this mode of cultivation. attack of nematode and white fly was major concern in polyhouse- cultivation.

7

Sustainable Agriculture and Food Systems in Asia Pacific: Challenges and Opportunities

G.P. Reddy[1*], Shyam karan[2], Sayon Mukherjee[3], S. Siddartha Naik[4], M. Chenna Madhava[5], Nikhil J[5]

[1]Department of Agronomy, Institute of Agricultural Sciences, Banaras Hindu University,Varanasi, Uttar Pradesh
[2]ICAR-Indian Agricultural Research Institute, New Delhi
[3]Department of Soil science &Agril. Chemistry, Institute of Agricultural Sciences, Banaras Hindu University,Varanasi, Uttar Pradesh
[4]Department of Agronomy, College of Agriculture -Bapatla, ANGRAU Lam Guntur, Andhra Pradesh
[5]Department of Extension Education, Institute of Agricultural Sciences Banaras Hindu University,Varanasi, Uttar Pradesh

Abstract

In many parts of Asia and the Pacific, farming plays a vital role in the rural economy, contributing significantly to GDP and employment. To ensure the well-being of rural populations, promote decent income, and foster inclusive growth and poverty reduction, it is imperative to support viable farming systems, especially for smallholder farms. However, agricultural systems across Asia Pacific face multiple challenges, including climate change and rising energy costs. Sustainable and resource-efficient agricultural practices are crucial in helping farmers adapt to these changes, sustain their livelihoods, and mitigate greenhouse gas emissions. Additionally, sustainable food value chains are essential for innovative poverty alleviation pathways. Notably, food waste reduction is an often overlooked aspect of sustainable food systems. Globally, a substantial 33% of all harvested food is wasted. Addressing food waste presents a significant opportunity to enhance food security and minimize the environmental impact of agricultural systems throughout the supply chain. UN Environment Asia Pacific has actively engaged in the global Think Eat Save campaign to reduce food loss and waste. This paper explores the importance of sustainable agriculture and food systems in Asia Pacific, focusing on challenges and opportunities. It emphasizes the need for adaptive and resilient farming practices, sustainable food value chains, and the reduction of food waste to achieve sustainable food production systems. The ultimate goal is to implement resilient agricultural practices, enhance productivity, promote ecosystem preservation, build capacity for climate change adaptation, and progressively improve land and soil quality.

Keywords: *Sustainable food production, SDGs, Poverty, Climate smart agriculture.*

1. Introduction

The agriculture sector in Asia and the Pacific plays a vital role in the rural economy, making significant contributions to both GDP and employment. This sector serves as a cornerstone for food production, economic growth and poverty reduction in the region. Understanding the importance of farming in the region's rural economy, as well as the need for viable farming systems particularly for smallholder farms, is essential for sustainable development.

Importance of Farming in the Rural Economy

Farming has a substantial impact on the rural economy of Asia and the Pacific. It serves as a primary source of income and livelihood for rural populations, contributing significantly to the region's Gross Domestic Product (GDP). The agricultural sector encompasses a wide range of activities, including crop production, livestock farming, fisheries, and forestry, providing diverse opportunities for economic development.

According to the Food and Agriculture Organization (FAO), farming accounts for 29 percent of the region's GDP (FAO, 2021). This demonstrates the significant economic value that agriculture brings to the region, making it a crucial sector for sustainable development. The revenue generated from agricultural activities supports not only farmers but also other stakeholders in the value chain, including agro-industries, distributors, and exporters.

Farming Contribution to Employment

In addition to its economic significance, farming is a major source of employment in Asia and the Pacific. It provides livelihood opportunities for a large portion of the population, particularly in rural areas where alternative employment options may be limited. The International Labour Organization (ILO) reports that agriculture employs around 65 percent of the workforce in the region (ILO, 2020). This highlights the critical role of farming in reducing unemployment, poverty, and rural-urban migration.

Smallholder Farms and the Need for Viable Farming Systems

Smallholder farms form the backbone of agriculture in many parts of Asia and the Pacific. These farms are characterized by their small landholdings, limited access to resources, and reliance on traditional farming practices. Smallholders face various challenges, including climate change impacts, lack of access to markets and credit, and limited technological advancements. Ensuring the viability of farming systems for smallholders is crucial for their livelihoods and the overall sustainability of the agricultural sector.

To address these challenges, there is a need to support and promote the development of viable farming systems for smallholders. Sustainable agricultural practices offer solutions for enhancing the productivity, resilience, and profitability of smallholder farms. These practices include agroecology, conservation agriculture, precision farming, and diversified farming systems, among others. By adopting sustainable practices, smallholders can optimize resource utilization, mitigate environmental impacts, and improve their economic returns.

Farming plays a critical role in the rural economy of Asia and the Pacific, contributing significantly to GDP and providing employment opportunities. The sustainability and viability of farming systems, especially for smallholder farms, are essential for ensuring the long-term well-being of rural populations, promoting inclusive growth, and reducing poverty. By supporting sustainable farming practices and addressing the challenges faced by smallholders, we can enhance the resilience and productivity of the agricultural sector in the region, leading to sustainable development and improved livelihoods.

2. Climate Change and Threats to Agricultural Systems

Climate change poses significant challenges to agricultural systems in Asia Pacific, impacting productivity, sustainability, and food security. The region is particularly vulnerable to the effects of climate change due to its geographical diversity, high population density, and dependence on agriculture for livelihoods. Understanding the impact of climate change on agriculture, including increased energy costs and other challenges, highlights the need for adaptive and resilient farming practices.

Impact of Climate Change on Agriculture in Asia Pacific

Climate change has already begun to affect agricultural systems in Asia Pacific, with adverse impacts on crop yields, water availability, and livestock production. Rising temperatures, changing precipitation patterns, and increased frequency of extreme weather events, such as droughts, floods, and storms, disrupt agricultural activities and reduce crop productivity.

According to the Intergovernmental Panel on Climate Change (IPCC), the region is experiencing shifts in rainfall patterns, with increased variability and intensity of rainfall events in many areas (IPCC, 2019). This variability affects planting and harvesting schedules, leading to crop failures and reduced yields. Moreover, rising temperatures contribute to heat stress in crops and livestock, further impacting productivity.

Increased Energy Costs and Other Challenges

In addition to the direct impacts of climate change, agriculture in Asia Pacific faces challenges related to increased energy costs and resource limitations. As energy prices rise, farmers face higher costs for inputs such as fuel, fertilizers, and irrigation, impacting their profitability and sustainability. Limited access to affordable energy sources exacerbates these challenges, particularly for smallholder farmers with limited financial resources.

Water scarcity is another critical challenge for agricultural systems in the region. Changing rainfall patterns, combined with increasing water demand from urbanization and industrialization, pose risks to water availability for irrigation and other agricultural purposes. Without adequate water resources, farmers face difficulties in sustaining crop production and livestock management.

Need for Adaptive and Resilient Farming Practices

To address the challenges posed by climate change and other threats, adaptive and resilient farming practices are crucial. These practices enable farmers to cope with changing climatic conditions, reduce vulnerability, and maintain agricultural productivity.

Adaptive practices include the use of climate-resilient crop varieties, improved irrigation techniques, and efficient water management strategies. Diversification of crops and livestock can also enhance resilience by spreading risks and reducing dependence on single commodities. Furthermore, the integration of agroforestry and conservation agriculture practices helps maintain soil fertility, enhance water retention, and mitigate climate-related risks.

The adoption of climate-smart agriculture (CSA) approaches is gaining momentum in the region. CSA combines sustainable farming practices, climate change adaptation, and mitigation strategies to achieve multiple goals, including increased productivity, reduced greenhouse gas emissions, and improved resilience. Examples of CSA practices include agroecology, precision farming, and integrated pest management.

Climate change poses significant threats to agricultural systems in Asia Pacific, impacting productivity, sustainability, and food security. The region is witnessing adverse effects on crop yields, water availability, and livestock production due to changing climate patterns. Additionally, increased energy costs and resource limitations further challenge the viability of agricultural practices.

To address these challenges, adaptive and resilient farming practices are essential. These practices help farmers cope with climate change impacts, reduce vulnerability, and sustain agricultural productivity. The adoption of climate-smart agriculture approaches, coupled with efficient water management, crop diversification, and conservation practices, can enhance resilience and enable farmers to adapt to changing climatic conditions.

3. Sustainable Agricultural Practices for Adaptation and Livelihoods

Sustainable agricultural practices play a crucial role in enabling adaptation to climate change and ensuring the long-term viability of farming systems. These practices, characterized by their resource efficiency and environmental stewardship, provide multiple benefits to farmers and contribute to the resilience of agricultural systems. By reducing greenhouse gas emissions from farm activities, sustainable farming practices also contribute to climate change mitigation.

Role of Sustainable, Resource-Efficient Agricultural Practices

Sustainable agricultural practices encompass a range of techniques and approaches that aim to optimize resource utilization, minimize environmental impacts, and enhance productivity. These practices promote the efficient use of water, energy, and nutrients, while preserving soil health and biodiversity. Examples of sustainable practices include agroecology, conservation agriculture, organic farming, precision farming, and integrated pest management.

Agroecology, for instance, focuses on the ecological interactions and relationships within farming systems. It emphasizes the use of locally adapted crops, crop diversification, and the integration of livestock and agroforestry. Conservation agriculture, on the other hand, promotes minimum soil disturbance, crop residue retention, and diversified crop rotations to enhance soil health, water infiltration, and carbon sequestration.

Benefits of Sustainable Farming in Adapting to Change

Sustainable farming practices offer several benefits that help farmers adapt to the challenges posed by climate change and other threats. Firstly, these practices improve the resilience of farming systems by diversifying crops, enhancing soil fertility, and conserving water resources. By adopting sustainable practices, farmers can reduce their vulnerability to extreme weather events, such as droughts or floods, and minimize yield losses.

Moreover, sustainable practices contribute to the long-term sustainability of agricultural production. Through soil conservation, organic matter enrichment, and reduced use of agrochemicals, sustainable farming helps maintain soil health, nutrient cycling, and ecosystem services. This leads to improved crop yields, reduced input costs, and increased profitability for farmers.

Reducing Greenhouse Gas Emissions from Farm Activities

Agriculture is a significant contributor to greenhouse gas emissions, primarily through the release of nitrous oxide (N2O) from fertilizer application and methane (CH4) from livestock and rice cultivation. Sustainable farming practices offer opportunities for mitigating these emissions and reducing the carbon footprint of agriculture.

For example, precision farming techniques, including site-specific nutrient management and optimized irrigation practices, can significantly reduce the overuse of fertilizers and water, thereby reducing nitrogen-related emissions. Additionally, agroforestry systems, such as silvopasture or agrosilviculture, can sequester carbon in trees and woody biomass while providing additional economic and environmental benefits.

Furthermore, sustainable livestock management practices, such as improved feeding, manure management, and methane capture systems, can help reduce methane emissions from livestock production. By adopting these practices, farmers can contribute to climate change mitigation while maintaining productive and profitable farming systems.

Sustainable agricultural practices are essential for enabling adaptation to climate change and ensuring the livelihoods of farmers. By promoting resource efficiency, enhancing resilience, and reducing greenhouse gas emissions, sustainable farming practices provide multiple benefits. These practices contribute to the long-term sustainability of agriculture, improve farmer profitability, and contribute to climate change mitigation. Embracing sustainable farming practices is crucial for building adaptive and resilient agricultural systems that can thrive in the face of changing climatic conditions.

4. Sustainable Food Value Chains

The concept of sustainable food value chains emphasizes the need for a holistic approach that encompasses all stages of the food production and distribution process, from farm to fork. Sustainable food value chains aim to promote inclusive growth, enhance food security, and minimize environmental impacts. This section will discuss the importance of sustainable food value chains for inclusive growth, the potential for innovative pathways out of poverty, and the crucial role of addressing food waste.

Importance of Sustainable Food Value Chains for Inclusive Growth

Sustainable food value chains contribute to inclusive growth by fostering economic opportunities for various stakeholders, including smallholder farmers, agribusinesses, and rural communities. By adopting sustainable practices, such as environmentally friendly production methods, fair trade principles, and equitable market access, value chains can ensure that benefits are shared along the entire supply chain.

Inclusive growth within food value chains involves empowering smallholder farmers and marginalized actors by providing them with access to resources, information, and markets. This includes facilitating fair and transparent transactions, promoting capacity building and training programs, and integrating smallholders into formal markets. By creating inclusive value chains, the livelihoods of small-scale farmers can be improved, leading to poverty reduction and increased food security.

Innovative Pathways out of Poverty through Food Value Chains

Sustainable food value chains offer innovative pathways out of poverty by creating diverse economic opportunities, promoting entrepreneurship, and supporting value addition. By connecting smallholders with markets and providing them with incentives to adopt sustainable practices, value chains can enhance their income-generating potential and reduce their vulnerability to external shocks.

Value chain development often involves the integration of smallholder farmers into higher value markets, such as organic, specialty, or export markets. This integration can lead to improved access to finance, technology, and knowledge, enabling smallholders to increase their productivity and profitability. Additionally, value chain upgrading strategies, such as the development of processing facilities or the introduction of new product lines, can create employment opportunities and stimulate rural economic growth.

5. Addressing Food Waste as a Crucial Component of Sustainable Food Systems

Food waste reduction represents a crucial component of sustainable food systems within value chains. Globally, up to 33% of all food harvested is wasted, leading to significant economic, social, and environmental implications. Addressing food waste offers a major opportunity to improve food security, reduce environmental degradation, and enhance resource efficiency throughout the supply chain.

Sustainable food value chains can contribute to reducing food waste by implementing measures such as improved post-harvest handling, efficient storage and transportation, and enhanced market coordination. By minimizing losses at each stage of the value chain, from production to consumption, valuable resources such as land, water, energy, and labor can be conserved.

Initiatives like the UN Environment's Think Eat Save campaign focus on reducing food loss and waste along the entire food value chain. These initiatives aim to achieve targets such as halving per capita global food waste at the retail and consumer levels by 2030. By implementing sustainable food production systems, adopting resilient agricultural practices, and reducing post-harvest losses, food waste can be significantly reduced, contributing to a more sustainable and efficient food system.

Sustainable food value chains play a crucial role in promoting inclusive growth, poverty reduction, and environmental sustainability. By ensuring equitable market access, empowering smallholder farmers, and minimizing food waste, value chains can create economic opportunities, enhance food security, and conserve resources. Through innovative approaches, such as value chain upgrading and integration into higher value markets, smallholders can find pathways out of poverty and contribute to sustainable development.

FAO. (2019). Transforming food and agriculture to achieve the SDGs. Rome: Food and Agriculture Organization of the United Nations.

Food Waste Reduction for Improved Food Security

Food waste is a pressing global issue with significant implications for food security. This section will discuss global food waste statistics, its impact on food security, and the importance of minimizing the environmental footprint throughout the food supply chain. Additionally, the engagement in initiatives like the Think Eat Save campaign will be highlighted as a means to reduce food loss and waste.

Global Food Waste Statistics and Its Impact on Food Security

Globally, a staggering amount of food is wasted throughout the entire food supply chain. According to the Food and Agriculture Organization (FAO), approximately one-third of all food produced for human consumption, equivalent to 1.3 billion tonnes, is lost or wasted annually (FAO, 2019). This wastage occurs at various stages, including production, post-harvest handling, processing, distribution, and consumption.

The impact of food waste on food security cannot be understated. With millions of people worldwide suffering from hunger and malnutrition, reducing food waste presents an opportunity to address this issue. Food waste contributes to the inefficiency of the food system, squandering valuable resources such as water, land, energy, and labor. By reducing food waste, more food can be made available to feed the growing global population and alleviate food insecurity.

6. Minimizing Environmental Footprint throughout the Supply Chain

Food waste not only affects food security but also has a significant environmental impact. The production and disposal of wasted food contribute to greenhouse gas emissions, water pollution, and deforestation. When food is wasted, all the resources used in its production, including water, fertilizers, and energy, go to waste as well.

Minimizing the environmental footprint throughout the food supply chain is essential to address these issues. It involves implementing sustainable practices at each stage, from production to consumption. For instance, improving agricultural practices, such as precision farming, can optimize resource use, reduce chemical inputs, and increase yields. Efficient post-harvest handling and storage techniques can minimize losses and extend the shelf life of perishable foods.

Furthermore, sustainable packaging and transportation methods can reduce food spoilage during distribution. At the consumer level, raising awareness about proper food storage, portion control, and expiration date management can help reduce household food waste. By adopting these measures, the environmental impact of food waste can be significantly reduced, contributing to a more sustainable and resilient food system.

7. Engagement in the Think Eat Save Campaign to Reduce Food Loss and Waste

The Think Eat Save campaign, initiated by the United Nations Environment Programme (UNEP), aims to raise awareness and mobilize actions to reduce food loss and waste. The campaign encourages various stakeholders, including governments, businesses, and consumers, to take proactive steps towards reducing food waste throughout the supply chain.

Through the Think Eat Save campaign, efforts are focused on multiple levels. These include improving agricultural practices, optimizing food processing and packaging, enhancing distribution and logistics, and promoting responsible consumption habits. The campaign also emphasizes the importance of partnerships and collaboration among stakeholders to drive collective action and achieve meaningful impact.

By 2030, the campaign aims to halve per capita global food waste at the retail and consumer levels and reduce food losses along production and supply chains, including post-harvest losses (UNEP, n.d.). The Think Eat Save campaign serves as a platform to exchange knowledge, share best practices, and implement innovative solutions to tackle the global food waste challenge.

Food waste reduction is crucial for improving food security, minimizing the environmental footprint, and creating a more sustainable food system. The staggering amount of food wasted globally highlights the need for collective action at all levels of the food supply chain. By adopting sustainable practices, raising awareness, and engaging in initiatives like the Think Eat Save campaign, significant progress can be made in reducing food loss and waste, ensuring a more resilient and equitable food future for all.

8. Goals for Sustainable Food Production and Waste Reduction

Sustainable food production and waste reduction are essential for achieving a more resilient and environmentally friendly food system. This section will discuss the goals set by UN Environment Asia Pacific, the importance of reducing food losses along production and supply chains, and the implementation of resilient agricultural practices and sustainable food production systems.

UN Environment Asia Pacific's Target to Halve Per Capita Global Food Waste

UN Environment Asia Pacific has set an ambitious target to halve per capita global food waste at the retail and consumer levels by implementing sustainable food production systems and reducing food losses along production and supply chains (UNEP, n.d.). This target is aligned with the broader objective of the United Nations Sustainable Development Goals (SDGs) to ensure sustainable consumption and production patterns.

Reducing Food Losses along Production and Supply Chains

Food losses occur at various stages of the production and supply chain, including post-harvest handling, processing, storage, transportation, and retail. These losses not only lead to economic inefficiencies but also contribute to food insecurity and environmental degradation. Reducing food losses is crucial for maximizing food availability, minimizing resource waste, and improving food security.

To address this issue, interventions are needed throughout the food system. For example, improving post-harvest technologies and storage facilities can minimize losses due to spoilage, pests, and inadequate handling practices. Enhancing transportation and logistics can reduce delays and prevent damage to perishable goods. Additionally, raising awareness and providing training to farmers, processors, and retailers can improve knowledge and skills related to proper handling, storage, and quality control.

Implementing Resilient Agricultural Practices and Sustainable Food Production Systems

Resilient agricultural practices and sustainable food production systems are fundamental for achieving sustainable food production and waste reduction. These practices aim to increase productivity, minimize environmental impacts, and enhance the resilience of farming systems to climate change and other threats.

Resilient agricultural practices include climate-smart agriculture, agroecology, precision farming, and integrated pest management. Climate-smart agriculture focuses on adapting to and mitigating the impacts of climate change through practices such as crop diversification, conservation agriculture, and efficient water management (FAO, 2020). Agroecology emphasizes the integration of ecological principles into agricultural systems to enhance biodiversity, soil fertility, and pest control (Pretty et al., 2011). Precision farming utilizes advanced technologies to optimize resource use, reduce input waste, and improve yields. Integrated pest management promotes the use of natural pest control methods to minimize chemical inputs.

Sustainable food production systems encompass a holistic approach that considers the environmental, social, and economic dimensions of food production. These systems prioritize resource efficiency, biodiversity conservation, and equitable distribution of benefits. Sustainable practices include organic farming, permaculture, regenerative agriculture, and agroforestry. These practices focus on reducing chemical inputs, improving soil health, promoting biodiversity, and fostering ecosystem services.

By implementing resilient agricultural practices and sustainable food production systems, farmers can enhance their adaptive capacity to climate change, increase productivity, reduce environmental impacts, and contribute to food security and waste reduction.

Goals for sustainable food production and waste reduction, as set by UN Environment Asia Pacific, emphasize the need for concerted efforts to address food waste along the entire supply chain. Reducing food losses, implementing resilient agricultural practices, and adopting sustainable food production systems are essential for achieving these goals. Through the integration of sustainable practices and the promotion of responsible consumption, we can build a more resilient, efficient, and equitable food system that supports both human well-being and environmental sustainability.

9. Enhancing Resilience and Improving Land and Soil Quality

Enhancing resilience and improving land and soil quality are critical aspects of sustainable agricultural systems. This section will delve into the importance of strengthening capacity for adaptation to climate change and extreme weather events, as well as promoting sustainable land and soil management practices.

Strengthening Capacity for Adaptation to Climate Change and Extreme Weather Events

Climate change poses significant challenges to agricultural systems, particularly in the Asia Pacific region. Increasing temperatures, changing rainfall patterns, and more frequent extreme weather events such as droughts, floods, and storms have adverse impacts on agricultural productivity and food security.

To enhance resilience, it is crucial to strengthen the capacity of farmers and communities to adapt to these changing conditions. This involves providing them with the necessary knowledge, tools, and resources to make informed decisions and implement adaptive strategies. Capacity-building initiatives can include training programs on climate-smart agricultural practices, climate risk assessment, and the use of climate information and early warning systems. Access to appropriate technologies, improved water management, and diversified farming systems can also contribute to building resilience and reducing vulnerability to climate-related risks.

Promoting Sustainable Land and Soil Management Practices

Land and soil are valuable resources that play a vital role in agricultural productivity, environmental sustainability, and ecosystem resilience. Promoting sustainable land and soil management practices is crucial for preserving soil fertility, minimizing erosion, enhancing water retention, and mitigating the impacts of climate change.

Sustainable land management practices include conservation agriculture, agroforestry, terrace farming, and the use of cover crops. Conservation agriculture involves minimizing soil disturbance, maintaining soil cover, and diversifying crop rotations to improve soil health and reduce erosion (FAO, 2017). Agroforestry integrates trees into farming systems to enhance soil fertility, control erosion, and provide additional sources of income. Terrace farming, commonly practiced in hilly and mountainous regions, involves constructing terraces to prevent soil erosion and retain water. Cover crops, such as legumes or grasses planted between main crops, protect the soil from erosion, improve nutrient cycling, and enhance soil structure.

In addition to these practices, sustainable soil management involves the judicious use of fertilizers, efficient irrigation techniques, and the reduction of chemical inputs. Implementing precision agriculture, which employs technologies like remote sensing and GPS mapping, can optimize resource use and minimize soil degradation. Furthermore, organic farming practices, which avoid synthetic fertilizers and pesticides, contribute to soil health and biodiversity conservation.

By promoting sustainable land and soil management practices, we can protect the quality and fertility of soils, mitigate the impacts of climate change, and ensure the long-term productivity and resilience of agricultural systems.

Enhancing resilience and improving land and soil quality are integral to sustainable agricultural development. Strengthening capacity for adaptation to climate change and extreme weather events equips farmers with the necessary tools to navigate changing conditions. Promoting sustainable land and soil management practices safeguards soil health, minimizes erosion, and fosters sustainable agricultural production. By integrating these practices into agricultural systems, we can build resilience, mitigate environmental risks, and contribute to food security and sustainable development.

Conclusion

In this paper, we have explored various aspects of sustainable agriculture, focusing on the challenges and opportunities it presents. We have discussed the importance of farming in the rural economy of Asia and the Pacific, highlighting its significant contribution to GDP and employment, particularly for smallholder farms. Furthermore, we have examined the impact of climate change and other threats on agricultural systems in the region, emphasizing the need for adaptive and resilient farming practices.Sustainable agricultural practices have emerged as crucial solutions for adapting to change and reducing greenhouse gas emissions. By adopting resource-efficient techniques, farmers can not only sustain their livelihoods but also contribute to the overall resilience of the farming sector. We have also underscored the significance of sustainable food value chains in promoting inclusive growth and offering innovative pathways out of poverty.Addressing food waste has been recognized as a key component of sustainable food systems. The alarming global statistics of food waste highlight the need for concerted efforts to minimize waste and its impact on food security. Engaging in initiatives like the Think Eat Save campaign provides a platform to reduce food loss and waste along the entire food production chain, leading to improved food security and reduced environmental footprint.

In conclusion, sustainable agriculture offers immense opportunities for inclusive growth, poverty reduction, and environmental sustainability. It requires collaboration among various stakeholders, including governments, businesses, farmers, and consumers. Policy support, investment, and knowledge sharing are vital to drive the transformation towards sustainable food systems. By implementing sustainable practices, reducing food waste, and strengthening resilience, we can pave the way for a future where agriculture supports both the well-being of rural communities and the preservation of our ecosystems.

References

FAO. (2017).Conservation agriculture. Retrieved from http://www.fao.org/conservation-agriculture/en/

FAO. (2019). Transforming food and agriculture to achieve the SDGs. Rome: Food and Agriculture Organization of the United Nations. UNEP. (n.d.). Think Eat Save - Reduce Your Foodprint. Retrieved from https://www.unenvironment.org/thinkeatsave

FAO. (2020). Agroecology and the Sustainable Development Goals. Retrieved from http://www.fao.org/3/cb0368en/cb0368en.pdf

FAO. (2021). Agroforestry. Retrieved from http://www.fao.org/agroforestry/en/

FAO. (2021). Food loss and waste reduction: A regional overview. Retrieved from http://www.fao.org/food-loss-and-food-waste/regional-overviews/en/ UN Environment. (n.d.). Think Eat Save - Reduce Your Foodprint. Retrieved from https://www.unenvironment.org/thinkeatsave

Food and Agriculture Organization. (2020). Agroecology and the Sustainable Development Goals. Retrieved from http://www.fao.org/3/cb0368en/cb0368en.pdf

Food and Agriculture Organization. (2021). Asia and the Pacific: Key Statistics and Indicators. Retrieved from http://www.fao.org/asiapacific/facts-and-figures/key-statistics/en/

Intergovernmental Panel on Climate Change (IPCC). (2019). IPCC Special Report on Climate Change and Land. Retrieved from https://www.ipcc.ch/srccl/

International Labour Organization. (2020). World Employment and Social Outlook: Trends 2020. Retrieved from https://www.ilo.org/wcmsp5/groups/public/---dgreports/---dcomm/documents/briefingnote/wcms_734438.pdf

Pretty, J., Toulmin, C., & Williams, S. (2011). Sustainable intensification in African agriculture. International Journal of Agricultural Sustainability, 9(1), 5-24.

Smith, P., Bustamante, M., Ahammad, H., Clark, H., Dong, H., Elsiddig, E. A., &Tubiello, F. (2014). Agriculture, forestry and other land use (AFOLU). In Climate Change 2014: Mitigation of Climate Change. Contribution of Working Group III to the Fifth Assessment Report of the Intergovernmental Panel on Climate Change. Retrieved from https://www.ipcc.ch/site/assets/uploads/2018/03/WGIIIAR5_Chapter11_FINAL.pdf

UNEP. (2021). Think Eat Save - Reduce Your Foodprint. Retrieved from https://www.unenvironment.org/thinkeatsave

8

Windrow Vermicomposting for Sustainable Agriculture

Kololgi S.D. and Koti R.V.

Green Foundation, NGO, Dharwad, Karnataka

Introduction

Increasing use and rise in the cost of chemical fertilizers, pollution due to their application and decline in soil fertility and micro flora are worrying facts to the farming community. Increased awareness about organic produce among the consumers is another reason to find out an alternative to chemical fertilizers. Vermicompost production on farm and its use in crop production is one way of mitigating these problems. Among the different methods, Windrow method of vermicomposting is the production of compost by piling organic matter or biodegradable waste, such as animal manure and crop residues as long rows (Windrows) on the surface of soil under the shade of trees without any permanent structures and later releasing required quantity of earthworms in wet beds. This vermibed technology is simple and cost effective and sustainable to adopt.

The prototypes have been exhibited in the premises of Dairy unit, University of Agricultural Sciences, Dharwad since long and many farmers seeing these were of the opinion that technology was simple and easy to adopt using the farm generated biomaterial waste with no additional cost. During 2020-21, when farmers of remote villages Dharwad district were interacted, it was noticed that still the technology was not followed by the farmers.Therefore, a team of Green Foundation based in Dharwad made an attempt to reach the farmers of remote villages of Dharwad district to make aware of this technology. After interactions with different categories of farmers, a modest attempt has been made to identify small and large farmers of remote villages viz. Basavarsikoppa and Holtikoti of Dharwad district during 2021-22 to sensitize and demonstrate windrow method of vermicompost production. The results, experiences of the farmers and finally the impact of the technology are discussed here in this presentation.

Objective

To interact and sensitize the small and large farmers and demonstrate the Windrow method of vermicomposting and its importance in sustainable agriculture.

Material and Methods

The team visited two times and interacted with the farmers of Basavarsikoppa (KalagatagiTq) and Holtikoti (DharwadTq) of Dharwad district.The farmers of two villages were briefed on the simple windrow method of vermicomposting showing the posters and ready samples of vermicompost as well as earthworms. Among the farmers gathered two(*Case I* and *II*) were selected whose fields were on the main road convenient for other farmers to view the demonstrations.

- ***Case-I:*** Shri Venkanagouda Chennappagoudar, Small Farmer, Basavarsikoppa (KalagatagiTq.)17 Km from Dharwad.
- ***Size of the Bed:*** 12' Length× 4' Breadth × 4' Height Prepared on 27.08.2021
- ***Source of Earthworms:*** Two Kg of worms (*Eisenia foetida*) obtained from Dairy Unit, MARS, UAS, Dharwad.
- ***Letting of Worms into the wormi bed:*** 2 Kg worms were let in on 08.09.2021
- ***Harvest of vermicompost:***0.8 ton worth Rs. 10,000 in two and half month period. Realising the worthiness of the technology, he raised two more beds.
  - ***Case-II:***Shri Basavaraj Kori, Holtikoti, Large Farmer, (DharwadTq.) 23 Km from Dharwad on Haliyal road.
  - ***Size of the Bed:*** 12' Length× 3' Breadth × 3' Height,Prepared on 08.08.2021
  - ***Source of Earthworms:*** Two Kg of worms (*Eisenia foetida*) obtained from Dairy Unit, MARS, UAS., Dharwad
  - ***Letting of Worms into the wormi bed:*** 2 Kg worms were let in on 16.08.2021
  - ***Harvest of vermicompost:*** 0.6tonworthRs. 7,500 in three months.
  - ***Realizing this,*** farmer of Holtikoti also raised **Seven** more beds in next three months and regularly applying to sugarcane crop.

Results and Discussion

A small farmer Sri Venkanagouda Chennappagoudar, by adopting the windrow method of vermicompost production in Basavarsikoppa village (KalaghatagiTq) harvested 0.8 ton in two and half months of time using mainly the soybean, maize stalks waste and dung(Table 1).The test report(SLN Bio Tech lab, Dharwad) of vermicompost indicated that it had moisture content (19.3%), bulk density(0.86%), total organic carbon(19.35%), N(1.89%), P_20_5 (1.62%), K_20(1.38%) and C:N (10:1), pH(6.98) and EC (3.22 dsm^1). As regards to heavy metals, it contained Zn (38 ppm), Cu (68), Cr (26 ppm), Pb (57 ppm), As_20_3 (5 ppm), Cd(2.7 ppm), Ni (27 ppm), Hg(0.07 ppm). All these values were found within the safe limit (Ambade *et al.,* 2021). Farmer was made aware of this test report and explained about the importance of test report for its use and also for commercialization in future.

Table 1: Details of Initiation of Windrow Vermicompost Production in Basavarsikoppa

Size	**Material Added**	**Worms(Kg)**	**Time taken for harvest(Months)**	**Qty. Produced (Tons)**
12' × 4' × 4'	Maize & Soybean Stalks, Dung, Sheep & Poultry droppings	2.0	2.5	0.8

Use of Vermicompost in Soybean

The Farmer of Basavarsikoppa arranged to apply all the vermicompost harvested to soybean crop in summer season (Table 1). Prior to the application, he arranged to sieve the compost and brought the consistency for use in sowing through seed cum fertilizer drill. The Farmer earlier in kharif and rabi had sown soybean with application of chemical fertilizers and compared the growth and yield of these two seasons with summer soybean with application of vermicompost. Results of kharif and rabi soybean crop with application of chemical fertilizers when compared with summer soybean crop with application of vermicompost in Basavarsikoppa indicated that Soybean yield in summer with only vermicompost has given considerable satisfactory yield compared to rabi where chemical fertilizer was applied (Table-2). Hence, it is noticed that vermicompost has provided required nutrients to some extent for the Soybean during summer. The Farmer expressed that the irrigation frequency was also reduced with 7-9 days interval during summer as against 5-6 days interval during rabi. Hence, the water holding capacity of the soil has also increased with vermicompost application in summer.

Thus, the farmer has been convinced with the advantages of windrow vermicomposting as no much cost was involved, and has further raised two more vermi beds. Another interesting fact noticed by the farmer was that when backyard poultry birds of his farm were fed with earthworms were observed to put on more weight and fetched him additional profit. Farmer also expressed that the other farmers noticing his vermicompost method of production were eager to adopt this low cost technology of windrow vermicomposting on their farms.

Table 2: Comparison of Soybean Yield between different seasons with Vermicompost and Chemical Fertilizers in Basavarsikoppa during 2021-22

	Kharif	**Rabi**	**Summer**
Variety	JS-335	JS-335	JS-335
Fertilizer/ Vermicompost/ha	NPK as recommended	NPK as recommended	0.8 ton Vermicompost alone
Diseases	Infected with rust	Less disease	Nil
Shattering	High	High	Low
Water Requirement	Normal precipitation	5-6 Days Interval	7-9 Days Interval
Yield / ha (Qtl)	08	20	16
Cost of Production	High	High	Low

Table 3: Windrow Vermicompost Demonstration in Holtikoti during August 2021

Size of the pit	**Material Added**	**Worms(Kg)**	**Time taken for harvest (Months)**	**Qty. Produced (Tons)**
12' ×3' ×3'	Paddy straw & Dung	2.0	3 Months	0.6 ton

Table 4: Comparison of Sugarcane (Cv. Chandghad 10001) performance between Chemical Fertilizer and VermicompostApplication at Holtikoti during 2021-22

	Chemical Fertilizers	**Vermicompost**
Fertilizer/ha	DAP(50 kg) / Urea(200 kg)/ Potash(25 kg)	Vermicompost(2.5 tons)
Weed intensity	More	Less
No. of side tillers at 60 days after planting	4-5	8-9
Crop foliage appearance at peak growth stage and arrowing	Light Green leaves and Early drying atarrowing	Dark Greenleaves and Retention of greenness at arrowing
Appearance of arrowing after planting	9 months	10 months
Irrigation interval/frequency	8 days	15 days
Cane yield/ha	60 tons	75 tons
Soil status	Hard	Soft and Porous

Another sugarcane farmer of Holtikoti raised Vermicompost bed under the regular guidance (Table-3) and harvested 0.6 ton of vermicompost in three months of time. He too was convinced about the ease of windrow method of vermicomposting and further raised seven more beds and harvested 3 tons in next three months. The Farmer applied 2.5 tons vermicompost/ha to the sugarcane at the time of planting (Table-4) and observed less intensity of weeds in vermicompost field than in chemical fertilizer field. He has regularly observed that vermicompost applied sugarcane appeared dark green with 40-50% more tillering. He also noticed that arrowing occurred one month later in vermicompost applied field than in chemical fertilizer. In addition, leaves remained more green at the time of arrowing in vermicompost applied field which was good source of green fodder for the cattle. The maintenance of more greenness at the time of arrowing in vermicopost applied field compared to chemical fertilizer was glaring, which attracted the attention of the many farmers. The irrigation interval in vermicompost applied field was increased by almost one week and thereby frequency of irrigations requirement was reduced, therefore farmer felt that there was increase in water use efficiency. The cane yield also has increased by 15 tons more in vermicompost (Table 4). Thus, the farmer has been convinced that the application of vermicompost has provided the nutrition to considerable extent if not on par with DAP/Urea to the sugarcane. The farmer of sugarcane crop expressed his experience with regards to land that land applied with vermicompost was soft and porous for field operations like ploughing saving the cost on fuel too. When Green Foundation team recently visited this farmer's field found that this famer has increased the production of vermicompost and regularly applying to the sugarcane thereby reducing the use of urea and DAP.

So, the demonstrations of windrow method of vermicomposting has not only realized the multiple benefits of vermicompost but also about the low cost involved in this technology of vermicompost production. Other farmers seeing the success of these two farmers are seen coming forward to adopt this technology.

Conclusion/Impact

These Farmers after realization, continued to produce vermicompost by windrow method on their farms and they also have come to known the multiple advantages of application of vermicompost, like good crop growth, increase in yield, improvement in soil fertility and soil tillage and increase in water saving. Both the demonstrated farmers want to take-up vermicompost production on commercial scale. This technology helps in doubling of the farmer's income along with improving the soil health and fertility.

No doubt if this technology can be popularized, it would be very good for sustainable agriculture also.

Reference

Ambade B., Sharma Y. and Sharma (2021). *Assessment of Windrow Composting Plant's Performance at Kern, Jodhpur, India.*
https://www.cabdirect.org/abstract/20163391324.

9

Sustainable Agronomic Technologies for Bundelkhand Region, Uttar Pradesh

***Nikhil Kumar Singh*[1], *U.P. Singh*[1], *Raghwendra Singh*[2] and *Amitesh Singh*[3]**

[1]*Institute of Agricultural Sciences, Banaras Hindu University, Varanasi*
[2]*ICAR-ATARI, Kanpur, Uttar Pradesh*
[3]*KVK, Kallipur, Varanasi, Uttar Pradesh*

Abstract

Agriculture is the bread and butter for more than 80% of the people of Bundelkhand region. Bundelkhand is majorly divided into two parts; Uttar Pradesh Bundelkhand (UPB) and Madhya Pradesh Bundelkhand (MPB). UPB consists of 7 districts whereas MPB consists of 6 districts of Uttar Pradesh and Madhya Pradesh, respectively. In UPB area, agriculture is rainfed in nature and successfully taken only one crop during rabi season regularly without fail due to extreme events. Moreover, agriculture in Bundelkhand is subsistence nature for the farmers. A crop failure due to aberrant weather is quite common in the region. Extreme climate events like, drought, flood, terminal heats etc. are common reasons for crop failure as well as anna pratha and farmers perish. The situation gets severe due to unavailability of modern promising sustainable agronomic practices and lack of diversification among the farmers for different crops and innovative ideas. This situation worsens due to the poor background of small and marginal farmers who accounted for about 84%. Furthermore, Bundelkhand is the region where crop choice of the farmers may change in a decade due to continuous crop failure. The present scenario of the region insists the scientific community to accept the challenge and turn this severe issue into a land of opportunity with the help of sustainable agronomic technologies.

Agriculture in Uttar Pradesh Bundelkhand (UPB)

Agriculture continues to be the principal occupation of the Bundelkhand people. Despite the low return from farming, people are engaged in agriculture. The region faces frequent drought and flood during the Kharif season resulting in converting many areas into fallow land. Major crops grown in UPB are cereals (wheat), pulses (gram) and oilseed (Mustard, linseed, and sesame) by most of the farmers. The traditional cropping technology and flood irrigation followed by the farmers in this region deepens the problem of the low profit and hardly matches the cost of cultivation from their income. This is the reason for mass migration of the young people from Bundelkhand to other part of UP and India. Whereas current trend shows that diversity is an encouraging trend and reduces dependence on a limited number of crops and crop mixes. Eventually, many farmers successfully begun to grow a number of crops like sesamum, rapeseed/ mustard, blackgram, pearlmillet, pigeon pea, maize, and wheat etc. Furthermore, rice caught the different trend and coverage make it major crop in kharif season. Farmers inclined towards specific variety and gaining good profits (eg. PB-1121 of rice).

Modern agriculture has created a shift to cereals and some selected oilseeds. In the past, farming in Bundelkhand was based on sustainable practices such as mixed cropping and staggered sowing. The choice of wheat was drawn from a basket of drought-resistant crops such as coarse cereals, millets and mixed-use grain and fodder. The region was known for coarse cereals, oilseeds and pulses, which were sustainable in the water-challenged zone and were cheaper to plant and grow. Over time, prompted by Government interventions, the farmers have changed to faulty new techniques and adopted new crops such as sugarcane, sunflower, castor, mentha, etc. which require chemical fertilizers, pesticides and a substantial amount of water.

Integration of policy makers and scientific fraternity may propose several countermeasures to the major concern to the agriculture. Furthermore, agriculture can help toward fighting against child malnutrition. Nutritive meals are a dream for small and marginal farming families in the region where 75-80 % of the population comes under this category. Malnutrition is another big challenge for the region to poor families. More than 80% of the population in Bundelkhand was suffering from anemia and energy loss (HDR, 2012). Cereals are staple food in low-income region of Bundelkhand, where they may contribute 55% of the dietary energy. Wheat and rice feeds almost every person and meets their energy and protein needs. In this article, an attempt has been made to cover the brief agronomical management to address many serious concerns of the agriculture faced by farming community of the Uttar Pradesh Bundelkhand (UPB) region.

Uttar Pradesh Bundelkhand (UPB)

UPB consist of seven districts Banda, Mahoba, Hamirpur, Jalaun, Jhansi, Lalitpur and chitrakoot and adjoin to the boundary of Madhya Pradesh. This region had a very ancient background of agriculture and allied occupation.

Beetelvine cultivation in mahoba, field pea in Jalaun and kathiya wheat is the signatures of the vast heritage of the region.

The irrigated area is 40 percent of the cropped land, crop productivity is low (although it has increased in recent years, yet it is below the averages in UP and MP, and much lower than the average productivity of the leading states), mechanization in agriculture is low, land holdings are small (average of two acres or so), inequality exists in land holdings, strong feudal land relations keep a stranglehold on farm labour. All these factors put together contribute to the low output in agriculture. Even with the increase in productivity of crops in recent times, it has not been sufficient, given the increase in input costs, to make agriculture profitable. The per person productivity and per hectare output of agriculture are both extremely low, and it appears that both poverty and stagnation in agriculture cannot be escaped unless per capita and per hectare productivity and profitability increase substantially. Many people are leaving cultivation for labour. Small farms are unsustainable and consequently more people turn to casual labour, to leasing land to sustain themselves, and their livelihoods have become peripatetic and precarious. The recent years of drought, with an ineffective drought mitigation system, with increasing debt burden have contributed to large-scale migration, which even though declining in recent years, has had an impact on the region. Families across the region have developed their own survival and growth strategies, which have involved mixed livelihood options, migration, and dependency on local feudatories (HDR, 2012).

Livestock management is a supplementary activity to agriculture and except for some cases in goat rearing (in which traditional goat rearing families are involved), usually a few animals are kept for self-consumption or small sales to organized and unorganized markets. There are some areas where livestock management is flourishing, such as goat rearing in the uplands and poultry in and around Jhansi. UPB district information is as followed:

Table 1: Coordinates and area of different districts of Bundelkhand area of UP

S No	Districts	Latitude	Longitude	Area
1	Banda	25^0 20' N	80^0 22' E	4413 Km^2
2	Hamirpur	25^0 58' N	80^0 12'E	4121.9 Km^2
3	Jalaun	26^0 08' N	79^0 23' E	4565 Km^2
4	Jhansi	25^0 27' N	78^0 37' E	5024 Km^2
5	Lalitpur	24^0 22' N	78^0 28' E	5039 Km^2
6	Mahoba	25^0 18' N	79^0 55' E	2884 Km^2

S No	Districts	Latitude	Longitude	Area
7	Chitrakoot	25^0 17' N	80^0 86' E	3216 km^2
	Bundelkhand (UP)			29262.9 Km^2

Major Agronomic Issues of the UPB

- Low productivity
- Unavailability of quality seed material
- Less exposure towards modern technologies
- Less irrigation facilities
- Injudicious use of available water
- Low nutrient use efficiency
- Lack of good agriculture machinery
- Rolling topography
- Depletion on ground water
- More than 80% farmers are small and marginal
- Malnutrition
- Migration of youth out of the Bundelkhand
- Less job opportunity
- Lack of skill labor
- Anna Pratha

The above issues are main culprits to create havoc among the farming community. Many problems can be addressed by the integration of technologies and the leadership of the institutional involvement with the farmers. NGO and government can start awareness programme in different steps but the role of agricultural universities like Banda University of Agriculture and Technology (BUAT), Banda and Central Agricultural University (CAU), Jhansi is pivotal. These institutions can collaborate with different stakeholders and train the farmers or farming society to upgrade the agriculture system and promote the farmers' livelihood with proper demonstration of the technologies. Close interaction and alliances with farmer societies and other institutions and stakeholders help farmers to regain their belief in sustainable technologies and resource use efficiency. Exposure of the farmer to world class technologies may be the oxygen to restore their faith in the agriculture business model. A common slogan is trending here in bundelkhand is the region is by default organic. Organic agricultural practices can take shape to flourish and show the newer dimension of the business model to the region.

To tackle the key issues related to agriculture may be solved using modern sustainable agronomic technologies and increase the use efficiency of the whole farming household. The area is untouched by innovative agronomic technologies. The following technologies we can use to ensure sustainable crop production.

Modern Technologies (Agronomic) for Sustainable Production

1. Development of Quality Seed Material

In UPB low productivity across every crop is common. Farmers are using age old seed material for normal production. On the other hand, availability of good quality seed always was the main constraints. Even the substandard private companies capture the agriculture market and sell their seeds in the region. State government departments are not able to meet the need of quality seed of the farmers. Furthermore, farmers had to travel far to collect good seed material, which is not possible for small and marginal farmers. Eventually, many areas can be easily found that farmers are using grains as a seed material. It is common among small and marginal farmers. Apart from these issues in the last decade the establishment of two agricultural universities is a chance to get good seed material. However, BUAT having six Krishi Vigyan Kendras (KVK) equipped with seed hub project are started to serve the purpose. Other ICAR institutions like IGFRI and CAFRI were already serving the region and helping farming community. KVK's runs the CFLD and FLD programme under the umbrella of BUAT sponsored by ICAR for popularizing new promising varieties and technologies among the farmers. Seed production of sesame, blackgram, green gram, Pigeonpea and rice during kharif season and Mustard, gram, Lentil wheat etc. during rabi season already started in BUAT.

2. Direct Seeding Through Seed Drills

Agricultural machineries can a play important role to transform any agriculture in any area of the country. Modern seed drills have good metering delivering systems. Also, farmers can adjust according to the crop for different crop geometry. Except, Pigeonpea almost every pulse crop grown in UPB region follow the same row x row distance. So, one setting can be fixed for these crops. There is need to change the single box seed drills to double box seed drills to use optimum seed rate. Sowing crop seeds mix with fertilizers like DAP or granular SSP lower the chances of uniform germination. Resulted farmers are using 20 % higher seed rate for crop sowing. Even, sesame and mustard can be sown through these seed drills with the help of spurious material or using coarse sand or DAP. Zero till ferti seed drills are very effective to reduce the drudgery and seed cost along with better establishment of crop in very effective manner. Use of happy turbo seeder is an effective tool to address residue burning issue and reduces drudgery, cost and increasing profit for the farmers. Furthermore, residue management addresses the environmental concerns raised by residue burning. Plain land may use laser land leveler to level the field efficiently and can increase the water use efficiency of the crops. It provides some uniform slope which help naturally draining the excess rain water.

Seed mix with sand | Seed delivery into seed box | Prior inspection before operation | Seed drill sowing

Fig. 1: Chronology of sesame seed sowing through seed drill

3. Use of Agro Advisory Services

Digital agriculture is the need of the hour to deal with extreme events in the regions. Climate change severely affects the area, and precise agro advisory based management practices help the farmers to become less vulnerable, more adaptable which leads to climate resilient crop production and sustainable management. Government of India releases big extreme events whereas; State departments deliver the information with the help of IMD and other weather services along with state agricultural universities. These institutions advocate and release advisories from time to time to the farmers for better care of their crops. Several weather-based apps are available digitally that can be downloaded and pursued by the farmers to enable them to be updated about weather activities.

4. Short Duration Crops

In the modern era, agriculture is affected by many factors, among them climate change effect is quite visible in these areas. Uneven rainfall patterns, drought and flood threaten the very livelihood of farmers in the region. Narrow window of rainfall and lack of water conservation practices make situation worst and it resulted in *kharif* fallow. "Anna pratha" is one of many issues by which *kharif* fallow prevails in the region. Short duration varieties of different crops can do wonders in the region and can add enough food in the food basket. Short duration crops like Black gram, Green gram and cowpea can be taken as a contingent crop though fix nitrogen into the soil and provide good soil cover to avoid surface soil erosion. Extra short duration variety of Pigeonpea can assure the farm income in aberrant weather like drought. All these practices are followed by selected farmers only and need to popularise among rest of the farmers as well as discourage the "anna pratha" to support crop planning.

5. Nutrient Management

In crop production, nutrient management is one of the most important parts. Instead of DAP, the recommendation of to apply full required amount of phosphorus through granular Single Super Phosphate (SSP) & Zn at the time of sowing with the help of seed-cum-ferti drills. To avoid extra input losses and save money, as we all know DAP is quite costly but very popular among the farmers. Granular SSP is used as a substitute of DAP because the soil of the UPB is deficient in sulphur and Zinc. Whereas SSP contains 16-18% phosphorus, 12% sulphur and 21% calcium. The area is known for blanket application of fertilizers. Farmers generally use only Urea as source for nitrogen and DAP as a source of phosphorus. Farmers normally focus on only these two fertilizers and avoid using potash. Farmer's practice refers to the fact that, they are using seed and fertilizers together, it is one of the reasons that they are using double seed rate. They own a single box of traditional seed drills, in which they put seed and fertilizer together. As we know, fertilizer should be placed below the seed not over the seed, but in traditional drill there is no guarantee to place fertilizer below the seed. This way of uncertainty directly affects the uniform germination of the crop. Thus, farmers use high seed rate instead of required quantity.

6. Residue Management to Avoid Burning

Residue burning practices were carried out by the farmers to escape from drudgery created by the loose residue into the field. Residue management can be done easily too if farmers follow the conservation agriculture practices. According to the principle of CA; retention of crop residues on the soil surface can make a difference. These stubbles help to moderate the temperature of soil. Residue burning is prohibited. It enhances production, reduces soil cracks, reduce soil erosion, increase water use efficiency, maintain optimum temperature, save irrigation water, minimize weeds, enhance microclimate of the soil, increase population of friendly microbes etc. The use of straw management system (SMS) behind the combine harvester uniformly distributes the straw into the field, helping in reducing drudgery. Further, it leads to less burning of the loose residues. Later, seeding in the presence of residue can cause trouble but advanced seed drills like Happy Turbo Seeder and Double disc planter can resolve the issue.

Happy Turbo Seeder

Straw Management System

7. Zero Till Technology on Sloping Lands

UPB receives nearly 900 mm of rain annually. Of this 80% is received during the Kharif season. Thus, rice crops can be grown with rainwater as grown under assured irrigation. Many farmers generally grow Rice crop on the flat, levelled, bunded and lowland fields and irrigate by flooding using tube-well waters. Thus, they grow transplanted rice under submerged conditions. However, there are many areas in the UPB where landscape is sloping and it's not possible to keep the soil submerged uniformly. Moreover, on these rolling topographies; sesame, black gram, green gram and pigeon pea crops can be grown on the uneven slopes using sprinkler irrigation in zero till conditions. The direct dry seeding of rice can be performed wherever, labor is a problem and later the fields can be submerged after rains which helps in water saving, reduced input and tillage costs.

8. Raised Bed Planting Techniques

Raised bed technique is getting very popular as a good production technique. In this technique, crops are grown on the raised beds made using a tractor drawn raised bed planter. Every raised bed and furrow are 67 cm wide. The raised bed is 37 cm wide, and furrow is 30 cm wide. Furrow is used for irrigation and also to provide drainage during high intensity rains. In black soil, which is known as "mar" in Bundelkhand making beds is a challenging job. But the job can be made quite simple if soil can be sprinkled with water for half an hour or a little more to moisten the soil. This avoids clodding. The reshaping of beds should be done only in Rabi season for stability of beds and to reduce soils erosion during the Kharif season. According to plant geometry, two rows of Rice, wheat, green gram, black gram and soybean are sown on the sloping shoulders of the beds while one row of maize or pigeon pea is sown on a single bed. Though sesame cultivation is practiced in kharif season, many times due to high rainfall and excess moisture, sesame crops have failed since the last two-three seasons. Poor drainage or drainage congestion often results in reduced sesame yields on the farmers' fields. For better crop production, raised bed technique for soybean and other pulses in kharif season is recommended. In the raised bed technique, the furrow drains the excess water from the field and regulates the wind movement to restrict the anaerobic condition which helps better crop establishment and higher yield. The free air movement helps to contain the crop free from diseases and insect attack. Light interception is high resulting in good photosynthesis and good crop yield. It has also been noted that in raised beds, uneven rain and high wind speed does not create lodging in standing crops.

Raised Bed Maize

Raised Bed Wheat

9. Controlled Traffic

Controlled traffic (CT) is a practice of planting crops between the tractor tyres and leaving the Tyre lanes without planting. In controlled traffic although the seed rate is less but the yields are not affected. The traffic lanes when used time to time, they get compacted leaving the rest of the field soils soft / uncompacted. The tyre lanes are subsequently used for other operations such as weed control, fixing the irrigation lines for sprinklers. Better aeration and good sunlight interception increase yield but reduces disease and pests. Much of the time and energy we put into the soil is to undo the compaction damage we have caused by driving machines all over them. As machines are getting heavier and heavier, this damage is extending deeper and deeper into the soil profile. CT aims to confine soil compaction to the least possible area of permanent traffic lanes. Controlled traffic cuts fuel, labour and machinery costs by reducing soil damage. This makes farming simpler, more reliable and less time-consuming. It also delivers environmental benefits, such as reduced water run-off and soil erosion, improved fertilizer use efficiency and improved carbon sequestration.

CT Wheat

10. Furrow Irrigation and Sprinkler Irrigation

Irrigation of heavy soil is important. During monsoon season, average rainfall of UPB is 900 mm. If the rainfall is normal, there will be no need for irrigation in *Kharif* season, but if the dry spell prevails for more than 7 days, farmers should go for the contingency irrigation. In this area, farmers are commonly using the flood irrigations methods to irrigate crops. They used sprinkler delivery pipes to irrigate the crop fields and put raisers into the store for ease operation. However, good demonstration of sprinkler irrigation and little bit convincing by the extension personnel can save plenty of water for next crop.

Whereas raised-bed technique followed by farmers can use furrow irrigation. Reports say that alternate furrow irrigation saves water and enhances water use efficiency in the presence of residue. In case of excess water, the farmers can use these furrows as a drainage channel. In wheat season, this technique will provide moisture for longer duration in residue retention condition and resulted high yield.

Furrow Irrigation

11. Nutrient Management

In crop production, nutrient management is one of the most important parts. Granular SSP is used as a substitute of DAP because the soil in UPB is deficient in sulphur and SSP contains 16-18% phosphorus, 12% sulphur and 21% calcium. 150 kg/acre of SSP is required which cost 975/- Rs only whereas, DAP cost 1350/- Rs for a single bag (50 kg). The area is known for blanket application of fertilizers. Farmers use majorly DAP and some use urea for better recovery. Farmers normally focus on only these two fertilizers and avoid using potash. It is recommended that 150 kg N/ha for rice and wheat is used while 200 kg N/ha for maize for example. Farmer's practice refers to the fact that, they are using together seed and fertilizers, it is one of the reasons that they are using almost double seed rate of many crops. They own a single box of traditional seed drills, in which they put seed and fertilizer together. As we know, fertilizer should be placed below the seed not over the seed, but in traditional drill there is no guarantee to place fertilizer below the seed. This way of uncertainty directly affects the uniform germination of the crop. Thus, farmers use high seed rate instead of required quantity.

12. Integrated Farming System

The major producer and contributor of food grains to the food basket of the country are equipped with plenty of natural resources and good irrigation infrastructure. However, the whole country pressured with rising population and unplanned colonization lead to rapid fragmentation of land holdings and shrinkage in fertile lands. Less opportunity in agriculture and lower profits treat farming as a secondary profession resulted huge migration of rural youth to nearby towns and cities for employment. In India, more than 70 percent of the total population of the country still live in villages and mainly depend on agriculture and/or related enterprises. Marginal and small farmers constitute more than 84 percent of the 115 million operational holdings in India which are cultivating only 29 percent of the arable land. Most of them are resource poor and work in diverse, risk prone environments (Singh *et. al*., 2011). Further, small land holders remained underprivileged of benefits of advancements in the field of general agriculture and specifically in agronomy. While the major beneficiary are progressive resourceful farmers. It arouses the major concern in the livelihood of the small and marginal household families. After independence different farming system have been evolved independently and being practices by the farmers without underlying principles of utilizing residues and wastes of the cropping and animals resulting wastage of resources and leads to environmental pollution. The average income of the small farmers from cropping alone is hardly sufficient to sustain their family throughout the year. Animal husbandry has been a vital part of existing farming systems across the country.

In addition to that, animal husbandry also did not provide a stable income. Several issues like, low milk productive animals, less awareness to their vaccination, unavailability of all-around year green fodder, infertility in cattle's, high mortality rate in poultry and goats are the factors that cause uncertainty in this sector. The small and marginal farmers are prone to poverty and because of that they are hesitant to adopt new technologies. Farming system can be described as a resource management system to achieve economic and sustained production to meet diverse requirements of the farm households while resources and minimizing the environmental pollution (Lal and Millu, 1990; Singh and Kumar, 2019). Integrated farming system models developed in different parts of the country involving different farm enterprises eg. dairy, duckery, poultry, agroforestry, horticulture, apiary, pisciculture and plantation crops etc. along with crops, increases net profit significantly as compared to cropping alone. These IFS systems are more sustainable and employment generative can contribute to improve the food security and livelihood.

13. Biofortification

In view of widespread deficiencies of vitamin, A, Fe, I, and Zn in humans, the problem was first attended to by physicians and nutritionists. Several supplements are available containing Fe and vitamin A all over the country. Furthermore, National Institute of Nutrition, Hyderabad, India, has developed common salt fortified with Iodine and iron (Sesikaran and Ranganathan, 2009). On the other hand, nutritionist adopted an approach like cereals based traditional foods with vitamins and minerals either at home or in factories for products such as flour, bread, biscuits etc. There are several cooking oils marketed in India with the label of added Vitamin A. One successful example is also very popular that iodized salt is being used in India to overcome iodine deficiency. Many suggested that we improve dietary by adding more meat, green vegetables, dry fruits, and fruits.

The drawback of these approaches is that major beneficiary are urban people whereas, rural peoples left out in countries like India in general and UPB in specific. Improving dietary composition with foods richer in Fe, Zn, and vitamins requires money and in many developing countries including India, poor purchasing power of a large part of human population is a major issue (Swaminathan, 2002). Genetic biofortification and agronomic biofortification can lead the way to address malnutrition among poor people. Many crops and varieties developed and produced around the world having denser Zn, Fe and amino acids but there are problems in the acceptance of GM crops in several countries.

Therefore, agronomic biofortification is the economical and swiftest method for biofortification of cereal grains with Fe, Zn, or other micronutrients in developing areas like UPB, where cereals are the staple food. Agronomic biofortification is the only way to reach the poorest of the poor rural masses, who will never have money to buy mineral supplements, nor can they afford to improve the components of their diet by incorporating animal products. From the biofortification viewpoint, foliar application is better and requires lesser amount of Fe and Zn fertilizers than their soil application. When cultivars with grains denser in Fe and Zn are developed, adequate Fe and Zn fertilization will be necessary (Prasad et. al., 2014). The genetic and agronomic approaches are therefore complementary to each other and should progress in tandem. However, a better understanding of the various reactions that micro mineral nutrients undergo in soil and the mechanisms involved in their absorption and translocation in plants, especially to grains, is required. That is a full work for developing countries which may facilitates future aspect to the area. There is the list of Biofortified crops having different varieties released in the country as follows:

Table 2: Biofortified Varieties released in the country by ICAR

Crop	Varieties	Nutritive Value	Grain Yield	Maturity	Adapted Region	Developed By	Remarks	Released year
Rice	CR Dhan 310	10.3 % protein	45 q/ha	125 Days	Odisha, MP and UP	NRRI, Cuttack	Pure line variety	2016
	DRR Dhan 45	Zinc 22.6 ppm	50 q/ha	125-130 Days	Karntaka, TN, AP and Telangana	IIRR, Hyderabad	Pure line variety	2016
Wheat	WB 02	Zinc (42.0 ppm) and iron (40 ppm)	51.6 q/ha	142 Days	PB,HR, Delhi, Raj, West UP, Una District of HP, Tarai region of UK	IIWBR, Karnal	Pure line variety	2017
	HPBW 01	Zinc (40.6 ppm) and iron (40 ppm)	51.7 q/ha	141 Days	PB,HR, Delhi, Raj*, West UP*, Una District of HP, Tarai region of Uttarakhand	PAU, Ludhiana	Pure line variety	2017
Maize	QPM 9 (Pusa Vivek improved)	High provitamin-A (8.15 ppm), lysine (2.67%) and tryptophan (0.74%)	Grain yield: 55.9 q/ha [Northern Hills Zone (NHZ)] and 59.2 q/ha [Peninsular Zone (PZ)]	93 days (NHZ) and 83 days (PZ)	Kharif season in J&K, HP, UK (Hill region), North Eastern states, MH, Karnataka, AP, Telangana and TN	IARI, New Delhi	Country's first pro-vitamin-A rich Hybrid maize	2017
	Pusa HM 4 Improved	Contains 0.91% tryptophan and 3.62% lysine which	64.2 q/ha	87 days	Kharif season in Punjab, Haryana, Delhi, Uttarakhand (Plain), Uttar Pradesh (Western region)]	IARI, New Delhi	Hybrid	2017

Crop	Varieties	Nutritive Value	Grain Yield	Maturity	Adapted Region	Developed By	Remarks	Released year
	Pusa HM 8 improved	Tryptophan (1.06%) and lysine (4.18%)	62.6 q/ha	95 days	Maharashtra, Karnataka, Andhra Pradesh, Telangana, Tamil Nadu	IARI, New Delhi	Hybrid	2017
	Pusa HM9 Improved	0.68% tryptophan and 2.97% lysine	52.0 q/ha	89 days	Kharif season in Bihar, Jharkhand, Odisha, Uttar Pradesh (Eastern region) and WB	IARI, New Delhi	Hybrid	2017
Pearlmillet	HHB 299	iron(73.0 ppm) and zinc (41.0 ppm)	G. yield: 32.7 q/ha Dry fodder yield: 73.0 q/ha	81 days	Kharif season in Haryana, Rajasthan, Gujarat, Punjab, Delhi, Maharashtra and Tamil Nadu	CCS-Haryana Agricultural University, Hisar in collaboration with ICRISAT, Patancheru	Hybrid	2017
	AHB 1200	Iron (73.0 ppm)	G. yield: 32.0 q/ha Dry fodder yield: 70.0 q/ha	78days	Kharif season in Haryana, Rajasthan, Gujarat, Punjab, Delhi, Maharashtra and Tamil Nadu	Vasantrao Naik Marathwada Krishi Vidyapeeth, in collaboration with ICRISAT, Patancheru	Hybrid	2017
Lentil	Pusa Ageti Masoor	65.0 ppm iron	13.0 q/ha	100 days	UP, MP and CG	IARI, New Delhi	Suitable for rainfed condition	2017

Crop	Varieties	Nutritive Value	Grain Yield	Maturity	Adapted Region	Developed By	Remarks	Released year
Mustard	Pusa mustard 30	Low erucic acid (<2.0%)	Oil content :37.7% Seed yield: 18.2 q/ha	137 days	UP, UK, MP and Rajasthan	IARI, New Delhi	Pure line variety; Suitable for timely sown irrigated conditions	2013
	Pusa Double Zero Mustard 31	Low erucic acid (<2.0%) Glucosinolates (<30.0 ppm) in Seed meal	Oil content :41.0% Seed yield: 23.0 q/ha	142 days	Rajasthan (North and Western parts), Punjab, Haryana, Delhi, Western UP, Plains of J&K and HP	IARI, New Delhi	Country's first Canola Quality Indian mustard variety	2016
Cauliflower	Pusa Beta Keshari 1	Contains high β-carotene (8.0-10.0ppm)	Curd yield: 40.0-50.0 t/ha	--	Nation Capital Region of Delhi	IARI, New Delhi	Country's first Biofortified cauliflower	2015
Sweet Potato	Bhu Sona	High β-carotene (14.0 mg/100 g)	Tuber yield: 19.8 t/ha Dry matter: 27.0-29.0% Starch: 20.0% Total sugar: 2.0-2.4%	-	Odisha	ICAR-Central Tuber Crops Research Institute, Thiruvananthapuram, Kerala	Pure line variety	2017

Crop	Varieties	Nutritive Value	Grain Yield	Maturity	Adapted Region	Developed By	Remarks	Released year
	Bhu Krishna	High anthocyanin (90.0 mg/100g) content	Tuber yield: 18.0 t/ha Dry matter: 24-25.5% Starch: 19.5% Total sugar: 1.9-2.2%	-	Odisha	ICAR-Central Tuber Crops Research Institute, Thiruvananthapuram, Kerala	Pure line variety	2017
Pomegranate	Solapur lal	High iron (5.6-6.1 mg/100g), zinc (0.64-0.69 mg/100g) and vitamin C (19.4-19.8 mg/100 g) in fresh arils	Fruit yield: 23.0-27.0 t/ha	-	Semi-arid regions of the country	National Research Centre on Pomegranate, Pune	Hybrid	2017

*Excluding Kota and Udaipur region in Rajasthan, Jhansi Region in West UP (Yadav *et. al.*, 2017)

References

Lal, R., and Millu, F.P. (1990). Sustainable farming for tropics. In : Sustainable Agriculture: Issues and Prospective. Vol. I (Ed.) R.P. Singh, pp 69-89. Indian Society of Agronomy, IARI, New Delhi.

Prasad, R., Shivay Y. S., and Dinesh Kumar, (2014) Agronomic Biofortification of Cereal Grains with Iron and Zinc. Advances in Agronomy, Vol. 125, pp. 55-91.

Sesikaran, B., Ranganathan, S. (2009) Salt fortification for micronutrient security. In: Symposium on Nutrition Security for India—Issues and Way Forward. 3–4 August, 2009, New Delhi, Programme and Abstracts. Indian National Science Academy, New Delhi, pp. 15–16.

Singh, J.P., Gangwar, B., Pandey, D.K. and Kochewad, S.A. (2011 Integrated Farming System Model for Small Farm Holders of Western Plain Zone of Uttar Pradesh. PDFSR Bulletin No. 05, pp. 58. Project Directorate for Farming Systems Research, Modipuram, Meerut, India

Singh, N. K. and Kumar, V. (2019) Ekikritkrishipranalisamay ki maang. Bundelkhand Connect pp 40-42

Singh, N. K. and Singh, U. P. (2014) Crop establishment methods and weed management on growth and yield in dry direct-seeded rice. Indian Journal of Weed Science, 46(4): 308–313.

Swaminathan, M.S., (2002) Agricultural progress and nutritional security. Inaugural Address, In: 2nd International Agronomy Congress on Balancing food and Environmental Security—A Continuing Challenge, New Delhi, November 26–30, 22 pp.

Yadava D.K., P.R. Choudhury, Firoz Hossain and Dinesh Kumar (2017). Biofortified Varieties: Sustainable Way to Alleviate Malnutrition. Indian Council of Agricultural Research, New Delhi.

10

Assessment of Climate Change Effect on Large Cardamom Crop in East Sikkim: Farmers' Perception

Posibia M. Daya Ram, S.M. Feroze, Baldev Singh and Aditya Vikram Nirala

College of Agriculture, Central Agricultural University, Imphal, West Manipur

Abstract

The paper aims at analyzing the effect of climate change on large cardamom based on farmers' perception in East Sikkim district of Sikkim. Data for the study were collected from 114 farmers through personal interview, focus group discussions and participatory rural appraisal methods. Simple random sampling technique was used for selection of the households. Simple random sampling technique was used for selection of the households. Descriptive statistical measures such as frequency, mean, percentage, standard deviation, chi-square test of independence and binary logistic regression were used for analysis and interpretation of the data. The study found majority of the farmers observed increase in the summer temperature, rainfall quantity and duration, snowfall/frost intensity/duration, landslides while decrease in the winter temperature. These changes in the climatic variables were reported to have visible impacts on the cardamom cultivation. Farmers observed decrease in flowering, fruiting, life span, shade requirement of the crop and increase in the tillering which overall resulted in the decline in cardamom productivity in the study area.

Keywords: Climate, Large cardamom, Problem tree, PRA, Sikkim

Introduction

Large cardamom is an important cash crop cultivated in Sikkim. It is also cultivated in Darjeeling district of West Bengal and some parts of North-eastern state like Arunachal Pradesh, Nagaland and Manipur. It is a low volume, high-value non-perishable crop grown organicallyat different altitudes under the trees. The State produces about 4,970 tonnes annually from an area of 23,312 hectares (Spice board, 2021). Among the four districts of Sikkim, North and East Sikkim district contributes largest in terms of area and production of large cardamom (GoS, 2016). Off-late, climate change has become a major concern for the large cardamom growers as many researchers reported its contribution to the decline of its production. Sikkim, as Himalayan states, is endowed with varied agro-climate ranging from tropical in the South to alpine in the North. Shrestha *et al.* (2010) reported that warming in the Himalayas was between 0.15°C and 0.60°C per decade in the last three decades. Seetharam (2012) revealed that in Gangtok, there is a decreasing trend in maximum temperature while an increasing trend in minimum temperature. The study also depicted the decreasing trend of annual rainfall amount in the study area. In a recent study, Kumar *et al.* (2020) reported that the warming was more prominent in the last decades in Gangtok and Tadong station, better reflected in the minimum temperature with varied precipitation across the observed period from 1961-2017.

These changes in the climate have reported a decline in production of large cardamom in Sikkim and other visible changes have been observed in plant behavior (Gudade, 2014; Jamwal, 2018). Climate change induced changes in local weather are reported to have increased the incidence of diseases and pests which result in decrease of productivity of the crop in Sikkim (Sharma *et al.*, 2016). With these points under consideration, a social study was conducted on farmers' perceptions of climate change and its effects on cardamom cultivation in East Sikkim.

Methodology

The study was carried out in East Sikkim District of Sikkim during 2020-21. The district occupies an area of 560 sq. km. in the south-east corner of the state. The elevation of study area ranges from 1273 meters (Rakdong) to 1935 meters (Rolep). Multi-stage sampling method was adopted in which selection of State, District, sub-divisions and villages were done purposively while selection of farmers was done through simple random sampling method. Out of the four sub-divisions in East Sikkim, Gangtok and Rongli sub-divisions were selected purposively. From each sub-division, two cluster villages were selected purposively. A sample size of 114 farmers was selected from the cluster villages through simple random sampling method.

To identify the major cardamom growing areas and for selection of the respondent farmers, secondary data was collected from the District Horticultural Department, Village *Pradhan* and other concern authorities. Primary data were collected through personal interview using structured interview schedule. Focus group discussions (FGDs) and participatory rural appraisal (PRA) were conducted in order to have general idea about the study area and farmers' perception on climate change. Descriptive statistical tools such as frequency, mean, percentage and standard deviation were used for analysis and interpretation of the data and are presented in tables and figures in the results and discussion.

Results and Discussion

Farmers' Perceptions on Climate Change

The study surveyed farmer's perceptions on climate change based on their observation and experience in the past 10 years. It was found that about 94.74 per cent of the respondent-farmers perceived changes in the one or more climatic variables while 5.26 per cent of them did not perceived any changes in the climate. The detailed analysis of the farmer's perceptions on different climatic variables is presented in Table 1and Figure 1.

Table 1: Perception about the change in amount of different climatic variables

Sl. no.	Climatic Variables	Increase		Decrease		Not sure		No Change	
		(f)	(%)	(f)	(%)	(f)	(%)	(f)	(%)
1.	Summer temperature	96	84.21	-	-	-	-	18	15.79
2.	Winter temperature	28	24.56	56	49.12	2	1.75	28	24.56
3.	Rainfall quantity	82	71.93	5	4.39	7	6.14	20	17.54
4.	Unexpected/untimely rainfall	55	48.25	-	-	19	16.67	40	35.09
5.	Duration of rainy days	59	51.75	3	2.63	4	3.51	48	42.11
6.	Drought	3	2.63	1	0.88	2	1.75	108	94.74
7.	Flood	1	0.88	1	0.88	4	3.51	108	94.74
8.	Snowfall/frost intensity/ duration	29	25.44	7	6.14	3	2.63	75	65.79
9.	Landslides	32	28.07	6	5.26	2	1.75	74	64.91

Table 1 reveals that majority of the farmers (84.21%) perceived increase in the summer temperature while the rest 15.79 per cent farmers found no change in the summer temperature. Similarly, about 49.12 per cent of the farmers perceived decrease in the winter temperature in the last 10 years. However, 24.56 percent of the farmers perceived increase in the winter temperature. Similarly, majority of the farmers (75.44%) perceived increase in the quantity of the rainfall during the monsoon season in the last 10 years while 4.39 per cent reported decrease in the rainfall quantity. The finding is in contrast with the result of Rymbai (2016) who reported the rainfall quantity is decrease. It may be possible that the farmers were not able to differentiate between the number of rainy days and the quantity of rainfall. Similarly, majority of the farmers (51.75%) perceived increase in the duration of rainy days as depicted in the table. Also, about 48.25 per cent of the farmers reported there is increase in the unexpected/untimely rainfall in the last 10 years. About 25.44 per cent of the farmers have perceived increase in the snowfall/frost intensity/ duration but majority of the farmers (65.79%) perceived no changes in the last 10 years. Changes in extreme climatic events like drought and flood were not much observed by the farmers in the study area. The findings are similar with the findings of Ram and Feroze (2020). However, 28.07 per cent of the farmers reported that the landslide has increased due to heavy rainfall during the monsoon season. Figure 1 reveals that majority of the farmers (72.81%) perceived no changes in the arrival of the rainfall and snowfall/frost. However, 7.89 per cent of the farmers perceived late arrival of the rainfall and early arrival of snowfall/frost.

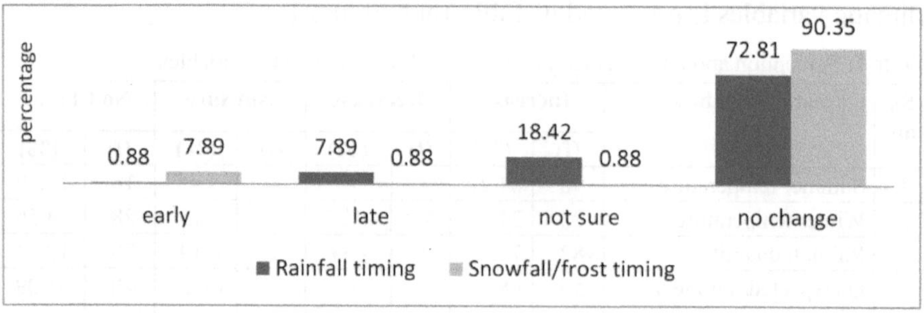

Fig. 1: Perception about the change in arrival of rainfall and snowfall/frost.

Community Perception on Various Climatic Events

The extreme climatic events in the villages were also captured through PRA by historical timeline mapping method and are presented in chronological order in the Table 3. In 1996 and 1997, farmers of Dalapchand village witnessed landslides which were associated with heavy rainfall. Also, *chirke* disease of cardamom was first reported in 2000 in this village. These villages used to experience hailstorms every year in the month of April. However, heavy hailstorms were witnessed in 2004 by farmers of Dalapchand and in 2017 by farmers of Rolep. The farmers of Tintek and Rakdong villages reported there was rainfall deficit during the monsoon season in 2016-17. In Rolep village summer temperature was reported to have increased since 2018 while winter temperature has decreased in 2020. Also, three out of the four cluster villages have perceived heavy rainfall in 2021.

Table 2: Historical timeline mapping of extreme climatic events

Year	Events	Village cluster I	Village cluster II	Village cluster III	Village cluster IV
1996-97	Landslides			✓	
2000	Started *chirke* disease of cardamom	✓			
2004	Heavy Hailstorm			✓	
2011	Earthquake			✓	
2012	Heavy Rainfall				✓
2016-17	Rainfall deficit	✓	✓		
2017	Heavy Hailstorm				✓
2018-21	High Temperature				✓
2020	Increase in coldness		✓		✓
2021	Heavy Rainfall	✓	✓		✓

Effect of Climate Change in Cardamom in the Last 10 Years

The changes in various climatic variables have affected the large cardamom in various ways which has visible impacts on its cultivation. Several farmers reported the decrease in cardamom yield due to several reasons in which climate change contributes directly or indirectly to the cause. Descriptive analysis of the visible changes observed by the farmers in large cardamom cultivation in the last 10 years is presented in Table 3.

Table 3: Changes in the cardamom in the last 10 years

Sl. no.	Particular	Increase (%)	Decrease (%)	Not sure (%)	No change (%)	Average change in productivity (%)
1	Flowering	-	22.81	1.75	75.44	10-20
2	Fruiting	0.88	71.05	0.88	27.19	30-40
3	Life span	-	81.58	-	18.42	50-70
4	Shade requirement	-	25.44	4.39	70.18	5-10
5	Tillering	41.23	-	3.51	55.26	30-40

Majority of the farmers (75.44%) found no change in the flowering of large cardamom in the last 10 years. However, 22.81 per cent of the farmers reported flowering of large cardamom had decreased which is associated with the reduction of the productivity of the crop. The average percentage of decrease in productivity was about 10-20 per cent. Although majority of the farmers found no change in flowering of the large cardamom, about 71.05 per cent of the farmers reported that the fruiting has decreased due to less rainfall during the flowering time. The decrease in fruiting has reduced the productivity of the crop by 30-40 per cent.Over 81.25 per cent farmers witnessed decrease in the life span of the crop. The economic life of the large cardamom was normally 12-15 years in the study area. However, it was found that the maximum life span of the crop in the study area was 7-8 years. The farmers reported that the yield of the crop decreases up to 50-70 per cent after 5-6 years of planting. The major cause of decrease in the life-span was mainly due to diseases and pests which indirectly or directly related with the change in climatic variables. The finding coincides with the findings of Gudade (2014) who reported climate change contribution to the decrease of economic life span of the large cardamom.

Large cardamom generally is a shade loving plant which grows mainly under the tree. The commonly use shade tree for growing cardamom in the study area are *Utis* (Alder), *Chillaune* (East Indian Almond), *Panisaj*, *Billaune*, *etc*. However, about 25.44 per cent of the farmers reported that the shade requirement in growing large cardamom has decreased that they could grow the crop in the open area. Over 41.23 per cent farmers found increase intillering (locally called as *Goghey*) which the farmers reported that diseases *viz*., *foorkey* and *chirke* could be the possible cause. The farmers also informed the excessive increase in tillering has also reduced in the fruiting thereby reduce the productivity of the crop up to 30-40 per cent.

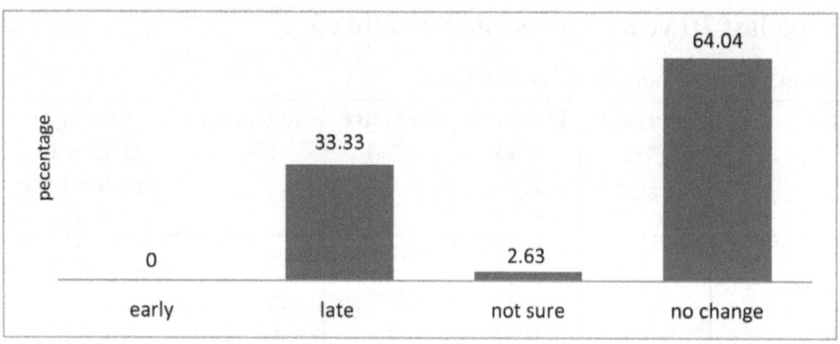

Fig. 2: Changes in the cardamom ripening over the last 10 years

Figure 2 depicts that over 64.04 per cent of the farmers found no change in the ripening of the fruits. However, 33.33 per cent of the farmers reported delay in the ripening of the fruits which could be associated with increase in the winter temperature. Also, delay in ripening had increased the fruit loss due to attack by the wild animals (rats, squirrels, porcupine) and was witnessed mainly in the upper hill area.

Table 4: Crop damage in Cardamom due to diseases and pests

Particulars	Damage (%)	Do you think it is due to climate change?					
		Yes (f)	Yes (%)	No (f)	No (%)	Not sure (f)	Not sure (%)
Diseases	40-50	80	70.18	4	3.51	30	26.32
Pests	10-20	16	21.62	38	51.35	20	27.03

*Crop damage:*The study witnessed crop damage caused by diseases and pests. Major diseases identified were '*foorkey*' and '*chirke*' disease of cardamom and the major pest identified were leaf caterpillar and small red ants. It is revealed from the table that the diseases had caused damage up to 40-50 per cent while pest up to 10-20 per cent. Also, majority of the farmers (70.18%) felt that the diseases were due to climate change while some of them claimed that it could be pollution due to increase in number of vehicles in the village. The finding coincides with the findings of other researchers (Brar *et al.*, 2017; Rijal,2014) who reported that climate change led to increase in the outbreak of diseases and the decrease in large cardamom production was due to *chirke* disease.

Problem Tree Analysis

In addition to the household survey, the existing problems faced by the large cardamom farmers were captured by conducting problem tree analysis with the villagers through PRA and FGDs and results are presented in Figure 3.

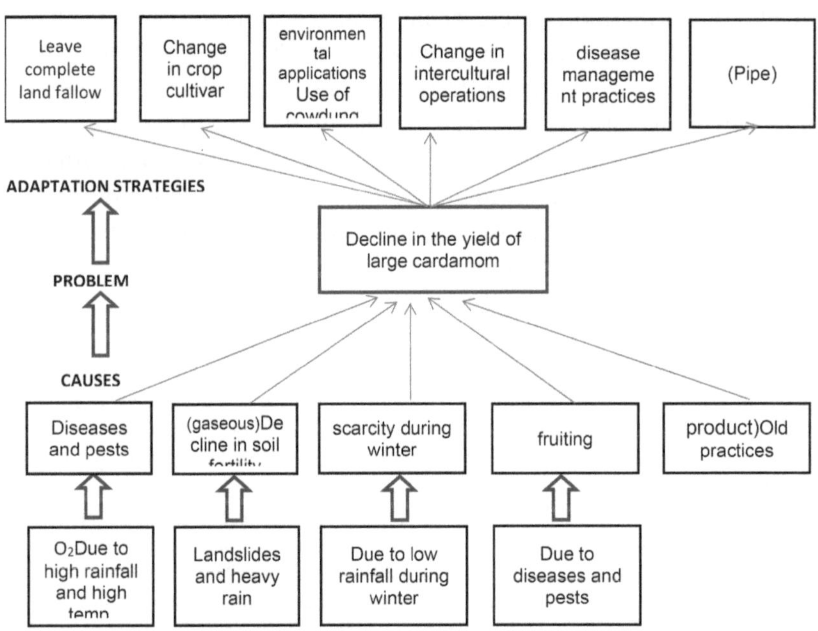

Fig. 3: Problem tree analysis on cardamom cultivation

The core problem identified was 'decline in the yield of large cardamom'. The primary causes of the problem were diseases and pests, decline in soil fertility, water scarcity during winter, less fruiting and use of old cultivation practices. The secondary causes were also generated from the primary causes. Heavy rainfall during monsoon season, high summer temperature, landslides, low rainfall during winter and diseases and pests were the secondary causes identified responsible for the primary cause. Overall it can be observed from the figure that changes in climatic variables were indirectly or directly related to the cause of the problem. Different adaptation strategies taken up by the farmers to cope up the problems were also identified. Leave land complete fallow, change in crop cultivar, use of cow dung and cow urine, change in intercultural operations, use of new diseases and pests management practices and irrigation during the dry season were the major adaptation strategies taken up by the farmers in the study area.

Conclusion

The study found majority of the farmers perceived the changes in the climate. They observed increase in the summer temperature, rainfall quantity and duration, snowfall/frost intensity/duration and landslides while they felt that winter temperature has decreased in the study area. The individual's perceptions were comparably similar with the findings with community perceptions from PRA and FGDs. Furthermore, these changes in the climatic variables have visible impacts on the large cardamom cultivation. The farmers in the study area observed decrease in flowering, fruiting, life span, shade requirement of the crop. However, the farmers observed increase in tillering of the crop. Crop damage due to diseases and pest were also observed by most of the farmers. Overall, these changes had decreased the productivity and thereby reduced the cultivating area. Therefore, the study suggests that actions need to be initiated to arrest the decline in large cardamom areas by finding out the solution to the disease infestation in large cardamom.

References

Brar, H.S., Sharma, A., Gill, J.S. and Sharma, S. (2017). Perception of Punjab Paddy Growers towards Climate Change. Indian Journal of Extension Education. 53(4): 80-83.

GoS.(2016). District wise data on large cardamom production.Available:http://www.sikkim.goi.in [April 8, 2021].

Gudade, B. (2014). Impacts of Climate Change on Large Cardamom (AmomumsubulatumRoxb.). In: Avasthe, R., Pradhan, Y. & Bhutia, K. (Eds.), Handbook of organic crop production in Sikkim,Gangtok: Sikkim Organic Mission, 2014, pp. 363-370.

Jamwal, N. (2018). Rise after the fall. Indian Himalayan climate adaptation programme-centre for media studies programme. Down to earth.Available: http://www.downtoearth.org.in/news/climate-change/rise-after-the -fall-61505 [April 7, 2021].

Kumar, P., Sharma, M.C., Saini, R. and Singh, G.K. (2020). Climate Variability In. Gangtok and Tadong weather observatories in Sikkim, India, during 1961-2017. Scientific Reports. Available:http://doi.org/10.1038/s41598-020-71163-ym[June 7, 2021].

Rijal, S.P. (2014). Impact of Climate Change on Large Cardamom-Based Livelihoods in Panchthar District, Nepal. The Third Pole: Journal of Geography Education, 13:38-38

Rymbai, D. (2016). Adaptation to climate change: An economic study of cereal crops in eastern Himalaya. Ph.D. Thesis (unpublished), Submitted to College of Post Graduate Studies, Central Agricultural University, (Imphal, Manipur) India.

Seetharam, K., (2012). Climate change synthetic scenario over Gangtok. In: Arrawatia, M.L. &Tambe M. (Eds.), Climate change in Sikkim patterns, impacts and initiatives, Information and Public Relations Department Government of Sikkim, 2012, pp. 1-18.

Sharma, G., Pratap, U., Dahal, D.R., Shama, D.P. and Sharma, E. (2016). Declining large cardamom production systems in the Sikkim Himalayas. Mount Research and Development, 36(3): 286-298.

Shrestha, B. B., Nakagawa, H., Kawaike, K., Baba, Y. and H. Zhang, H. (2010). Glacial lake outburst due to Moraine Dam failure by seepage and overtopping with impact of climate change. Annuals of Disaster Prevention Research Institute, Kyoto University. 53 (B): 569-582 Available: https://www.researchgate.net/publication/47528345_Glacial_Lake_Outburst_due_to_Moraine_Dam_Failure_by_Seepage_and_Overtopping_with_Impact_of_Climate_Change[Jan. 5, 2021].

Singh, R. and Feroze, S.M. (2020). Impact of Climate Change on Animal Husbandry: A Gender Perspective Study in Meghalaya. Indian Journal of Extension Education. 56(4): 31-36.

Spice Boardof India (2021). Major spice state wise area production 2021. Available: http://www.indianspices.com/sites/default/files/majorspicestatewise2021.pdf [Oct. 10, 2021].

11

Is it Biochar, the Green Engineered Material from Agri-waste? A Promise Towards Technology Transfer Supply Chain Leadership for Sustainable Development

Pradipta Chandra

Prestige Institute of Management and Research
Education and Health Sector, Scheme 54, Indore, Madhya Pradesh
Alumni, Indian Institute of Technology Kharagpur, West Bengal

Abstract

Biochar is a hard and solid material, generated from pyrolysis of lignocellulosic biomass like waste from agriculture, forest and wood residues, which is a unique material for soil amendments, environmental applications and additive for property enhancements of various polymeric materials for engineering applications. Theimportant properties related to surface area, pore volume and diameter, electrical conductivity (EC), cation exchange capacity (CEC), p^H, carbon, hydrogen, ratio of hydrogen to carbon (H/C), ratio of oxygen to carbon (O/C), ratio of carbon to nitrogen (C/N) aresignificant parameters of biochar for soil amendments. Despite its enormous potential, biochars have not been used widely due to unavailability of established supply chain with respect to the cost of production and standardization of technical parameters. Present review paper identifies the production process, benefits and features of the supply chain for the viable adoption of biochars.

Keywords: Biochar, Soil amendment, Agri-waste, Supply chain, Agricultural operations management, Technology transfer, Sustainable development, India.

1. Introduction

Biochar, produced from pyrolytic decomposition of lignocellulosic biomass, is a solid hard product and a unique green material for soil amendments. Its use has been identified in Amazon fertility as ancient indigenous practice. Through 'pyrolysis' (heating without oxygen) process ('pyro' which means caused by heat whereas 'lysis' means dis-integration), there are three types of yields we get: a solid product (biochar), a gaseous product (syngas/ synthesis gas), and a liquid (bio-oil). The solid item, biochar, is applied for soil amendments. It can be used in agricultural, industrial, and environmental as a functional material. Solid char, a common pyrolytic outcome is charcoal, typically used to designate the solid char product that is ignited for fuel, whereas "biochar" designates the unique solid char purposively used as a soil amendment agent. On the other hand, the gaseous and liquid counterparts of pyrolysis have commercial applications like green energy, feedstock raw materials, and precursors (Fig.1).

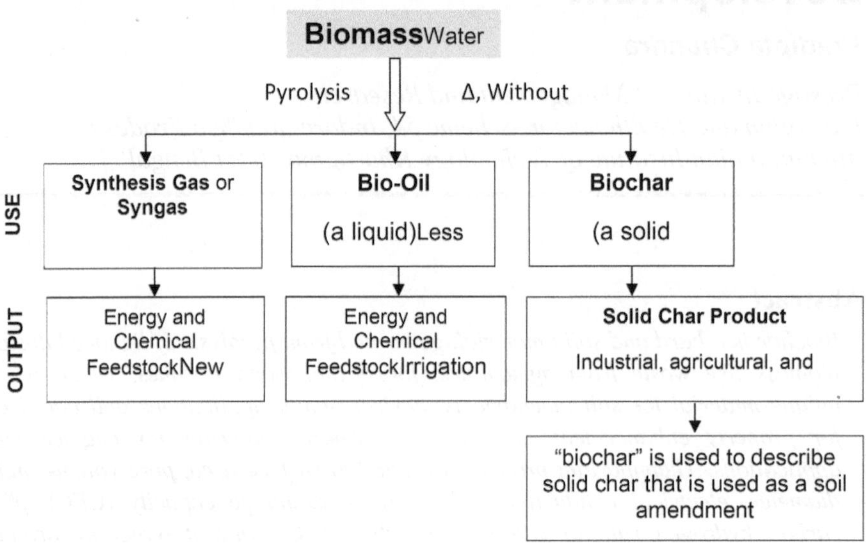

Fig. 1: Biomass to biochar conversion

Biochars are commonly used as a soil amendment agent although implication of biochars needs a holistic evaluation by considering it as a part of the pyrolysis system that coproduces biofuels. The contribution of biochar is note worthy:

- Biochars can be used in improvement of agricultural productivity (yield) through amendments of soil and intensification of input use efficiency.
- Biochars contribute positively in mitigating climate change through carbon sequestration.
- It supports fire prevention.

The benefits and impacts of biochar depend upon the crop and the biological and physical settings of application. Depending on the feedstock category and available pyrolysis technology biochar itself is a heterogeneous product. Conventional biochar systems follow fragmented supply chains. This is done by adopting simple pyrolysis process on locally available biomass by ignoring byproducts gas and liquid. Eventually, the complete economic benefit of the particular system is not realized and thus a modern biochar production system has not attained a competitive commercial success in spite of its potential to success in agricultural productivity,soil functioning, and carbon sequestration. However, the dynamic and long-term effect of biochar needs further study as it has been found that all biochars do not contribute in reduction of nitrogen (N) loss universally under all soil conditions. By applying biochar on neutral soil generally leads to a higher soil p^H, which support increase NH_3 volatilization. Hence, understanding the conditions where biochar could help reduce NH_3 volatilization is vital to propose the benefits of biochar applications.

2. Benefits of Biochar in Agricultural and Soil Ecosystem

Agricultural ecosystems can get diversified benefits from biochar like increased soil organic matter content, improved water retention, increased microbial activity, increased nutrient recycling and retention and finally reaches to improved yields (Fig.2).

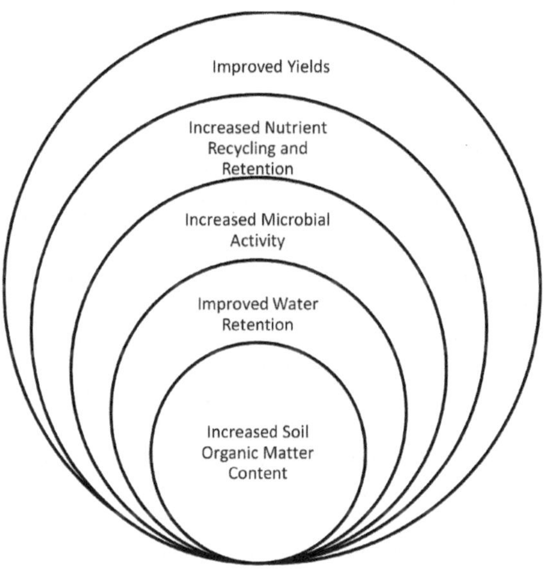

Fig. 2: Various Benefits of Biochar in Agricultural Ecosystem

Close-loop Waste Management

Biochar has the prospect in close-loop waste management by converting organic material (from industries, such as, agriculture and forestry) into low-cost soil amendments. The potential benefits of different biochar systems are elaborated below.

Individual benefits: Improved agricultural productivity through soil functioning

Controlling p^H of soil and cation exchange capacity

Longitudinal studies reveals that boosting yield happens mostly due to improvements in soil p^H. Biochar has two principal effects (Fig. 3, Table-1) on soil acidity, a common controlling factor for agricultural production.

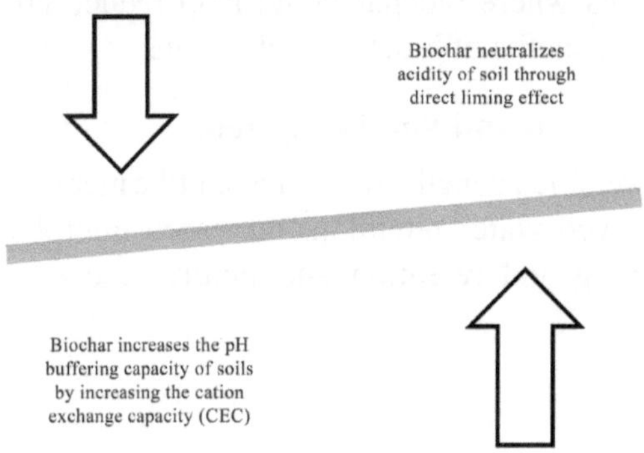

Fig. 3. Relationship of soil pH and cation exchange capacity

Table 1: Function of biochar as soil amendment agent

Sl. no.	Soil Function	Role of Biochar
01.	Liming Effect	Biochar helps in neutralization of soil acidity through a direct liming effect, improves the soil structure suitable for microbial communities, controls bioavailability and retention of nutrients, and leaching, and facilitates plant toxicity above or below certain thresholds.
02.	Buffering	Biochar increases buffering capacity of soils by controlling$p^H$and increasing the cation exchange capacity (CEC) (asnegative functional groups (COO^- and $O^-$are bound to the surfaces of biochar [89].
		Biochar enlarges the reactive surface area available area in low-fertility soils, which help increasing CEC and buffering capacity, water retention, increases porosity, to support plant roots more available soil volume.
		Increases in CEC,particularly in coarse-grain or highly weathered soils
03.	Both Liming and Buffering	For low fertility soils simultaneous effect of biochar's buffering and liming can improve crop productivity [4, 22, 66].

A global study indicates that biochar has its highest impacts on soil health when it is applied to low p^H, less cation exchange capacity, and soils having coarse-texture [20] (Table-2).

Table 2: Impact on crop yield on application of biochar

Sl. no.	Where	Parameter	Statistics	Reference
01.	Tropical with acidic soils	Yield increase	by 25% (average)	Jeffery *et al.*[40]
02.	Upland in China	biochar from rice straw	boosted by 17% (rapeseed production)	Jin *et al.* [41]
03.	Regions of tropical climate in sub-Saharan Africa (SSA)	increase in yield	Yield increase: 0.18 to 1.00 ton ha^{-1} $year^{-1}$	Dickinson *et al.*[23]
04.	Regions of temperate climate, Northwest Europe (NWE)	increase in yield	Yield increase: 0.07 to 0.28 ton ha^{-1} $year^{-1}$	Dickinson *et al.*[23]
05.	Poor soils in Labrador, Canada	forest biomass in beet field	Yield increase: 2.9 to 11.4 Mg ha^{-1}	

A comprehensive literature review provides a descriptive statistics on various benefits of biochar with respective to the crop yield improvements. Crop yields increases by 25% on average in tropical and acidic soils.On the upland of China, a field-trial of 5-year explores that rapeseed production has been boosted by 17% on application of biochar produced from rice straw. Another international study, conducted by Dickinson *et al.*, estimates that in the regions of tropical climate, a representative area in the sub-Saharan Africa (SSA), the crop yield increased from 0.18 to 1.00 ton ha^{-1} $year^{-1}$. On the other hand, the regions of temperate climate, in Northwest Europe (NWE), the similar yield is observed from 0.07 to 0.28 ton ha^{-1} $year^{-1}$. An unique study conducted in Labrador, Canada on the relatively weathered soils reveals that by applying biochar, produced from forest biomass, to a field boosted the yield from 2.9 to 11.4 Mg ha^{-1}.

Supply Retention of Nutrient

Application of biochar has been demonstrated for improving three fundamental soil nutrient management processes: nutrient supply, nutrient retention, and nutrient cycling (Fig. 4). After pyrolysis nutrients, retained in the mother feedstock biomass, are accumulated in biochar and same are added to soil under study.

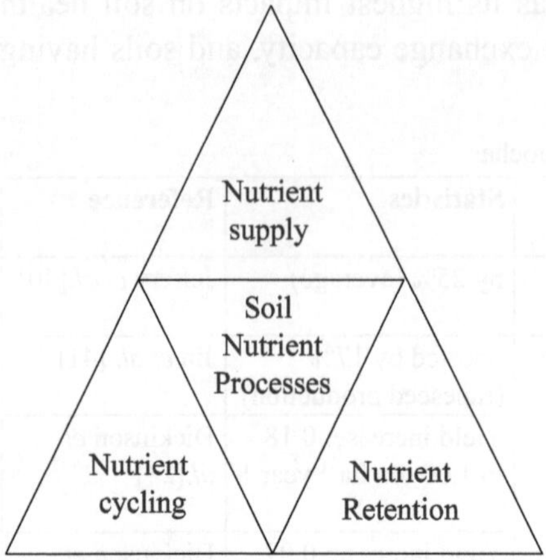

Fig. 4: Soil nutrient process by biochar

Nutrient Supply

Biochar has a direct fertilizing effect and facilitates nutrition growthinsoil, more particularly potassium (K) and phosphorus (P) although study reveals that forhardwood(feedstock of low-nutrient) and biochar having high-surface area (produced under higher pyrolysis temperatures) can have a complex effect on the bioavailability of soil nutrient stocks. Biochars made from low-ash feedstocks (hardwood) have lower concentration of P and K, but conventionally, a higher pyrolysis temperature increases P and K content on biochar. Thus, the contents of such nutrients within biochar will be governed by both, the quality/type of feedstock and conditions of pyrolysis (Fig. 5).

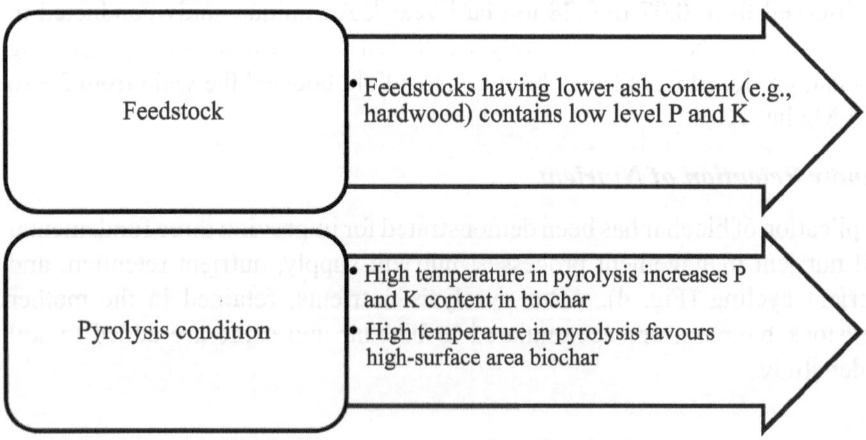

Fig. 5: Influence of feedstock and pyrolysis condition on the characteristics of biochar

Nutrient Retention

Besides supplying nutrients directly, biochar can help retain important valuable soil nutrients through decreasing losses due to leaching. Study says that nitrogen leaching is decreased in temperate soils due to the adsorption phenomena by the biochar. Another study, conducted by Randolph *et al.* on different biochars produced at various pyrolysis temperatures and rates, reveals that most of the feedstock biochars prevent NO^{-3}, K, Ca, and Mg leaching. Biochar has the property to save fertilizer through a reduction in leaching and volatilization of nitrogen. According to Gaunt and Lehmann [33] a reduction of loss of fertilizer by 10% to 30% is required for crops. Considering soil morphology, biochars having high-porosity and high-surface area biocharsare suitable to reduce nutrient leaching. In a contrary, there are adverse effects on nutrient suppliesin low-fertility soils. Following figure illustrates how specific biochar properties can be engineered for target uses.

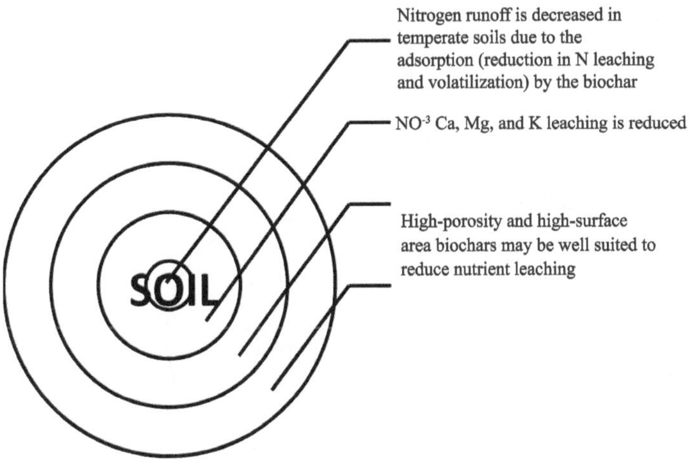

Fig. 6: Retaining important soil nutrients by decreasing losses from leaching

Nutrient Cycling

Biochar also monitors nutrient cycling in soils. Nitrogen cycling can be controlled by biochar application through variouis mechanisms, more particularly fixation biological N_2, ammonia (NH_3) volatilization, and denitrification. Agricultural operations which use urea-based fertilizers, apply higher quantity of synthetic nitrogen. Biochar application has demonstrated a reduction in NH_3 volatilization in above situation as well as in tropical, rocky, or otherwise low-fertility soils, by establishing it a favorable strategy for a varied set of farmers to enhance and intensify their nitrogen use efficiency.

Benefits to Compost Production by Co-composting

Biochar can be useful to improve the efficacy of composts. Suitable co-composting (through incubation of biochar with main compost) improves crop yield, nutrition supply and its retention, and biological activitysoil. In a study, it has been found that total N losses were reduced by 50% from a compost, generated from poultry litter, when it was amended with biochar. Biochar makes substantial improvements in compost production and application, primarily due to biochar's

- porous and hierarchical (meso)structure
- very highsurface area which allows nutrients, water, and organic matter to be adsorbed; hence, surface area is an significant measured parameter for biochar characterization.

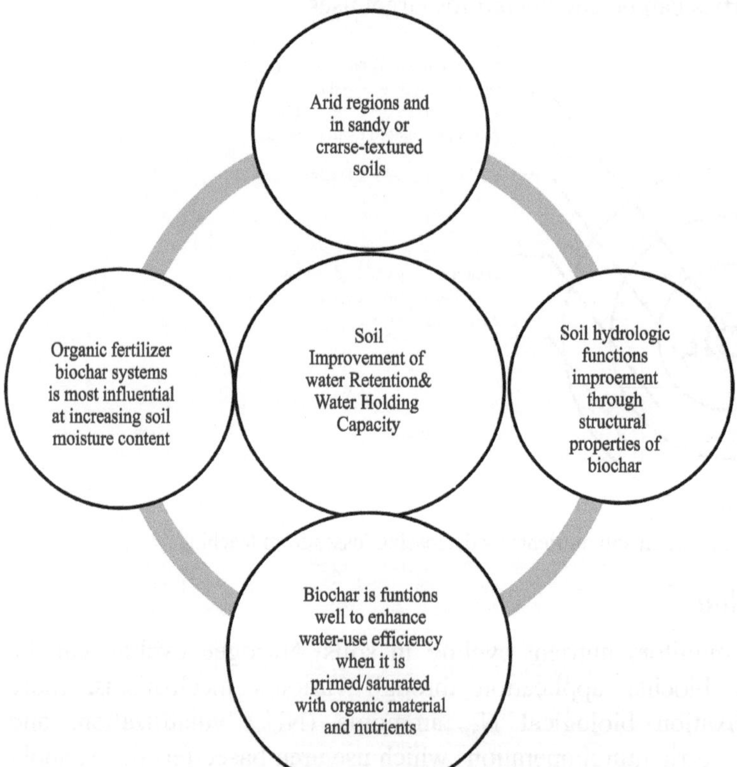

Fig. 7. Improvement of water retention and water holding capacity

Improvement of Water Retention and Holding Capacity

Biochar has demonstrated its capacity to improve an array of soil-water interactions, especially in dry regions, sandy or coarse-textured soils,. A recent study has demonstrated that the structural dimensions of biochar improve soil hydrological functions (Fig. 7) in the following context:

- average increase in aggregate stability by 8.2%
- increase in available water holding capacity by 15.1%
- increase in saturated hydraulic conductivity by 25.2%

Field experiments have also supported that biochar is perfect in improving water-use efficiency when it is suitably mixed with nutrients and organic material. A study on tropical soils, where a comparative study was conducted with a range of biochar with co-compost amendments, reveals that a co-composted biochar-fertilizer systems is most effective for increasing soil moisture content. It is important to note that the available water for plant may vary with the duration the biochar has been working in the soil.

Social Benefits Through Climate Change mitigation

As a promising carbon sequestration (CS) technology, biochar is projected to have a highest possibility to offset by predicting that all biomass everywhere in world can be transformed into biochar by adopting low-emission through slow pyrolysis, high-yield technology with maximizing bioenergy. Study also shows that biochar has a tremendous carbon sequestration potential with 0.7-1.3Gt CO_2. Benefits in climate change from biochar are comprehended predominantly through carbon sequestration in the field of agricultural. According to Woolf *et al.* carbon sequestration and off setting fossil fuel are two major contributors inreducing emissions. The advantage of fossil fuel balance remarkably depends on the comparative intensity of carbon in bio-energy and fossil fuel. Bioenergy proposes the highest reduced emissions provided coal is replaced (15%).

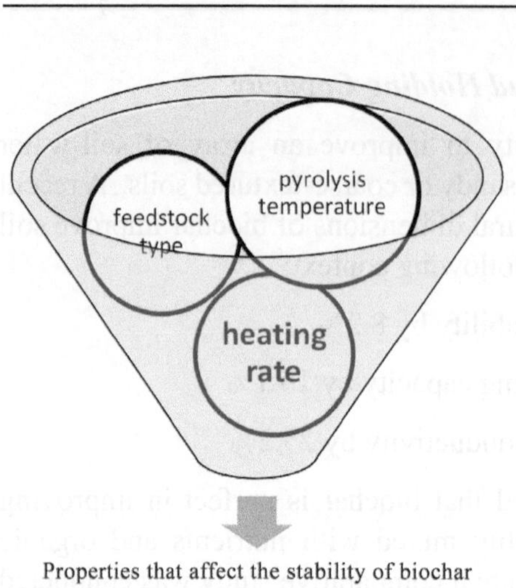

Fig. 8: Parameters affecting the stability of biochar

Carbon Sequestration

Biochar proposesa valuable technology for carbon storing as it can deferthe process of replenish carbon to air significantly once it is captured through photosynthesis. Approximately 10% to 15% of original carbon is being converted through pyrolysis of plant biomass with a highly concentrated and biochemically stable form. To understand its role as a carbon storage technology it is required to know the stability and breakdown pattern of biochar although differences in environmental conditions, different analytical methods, and versatile propertiers of biochar make it difficult to determine period of existence of biochar derived carbon in soils. The inherent stability of biochar carbon is a function of its properties and of the soil and environmental conditions where it is applied. The principal properties that affect the stability of biochar are influenced by feedstock type, pyrolysis temperature alongwith the rate of heating (Fig. 8). Any alteration with these production factors yield a wide range of physicochemical properties that specify differences in the stability of the aromatic carbon molecules present in char material.

O/C Ratio, H/C Ratio and negative emission technologies (NET) for sustainable development

Negative emission technologies (NET) is articulated in the low O/C ratio and decreased volatile matter relative to biochars by using low-temperature [78]. Aiming at limiting global temperature rise to 1.5 °C, it must include mass arrangement of negative emission technologies (NET) that capture carbon from the atmosphere. For evaluating the stability, volatile composition, and fixed carbon composition of biochar the O/C ratios, as well as the H/C ratios, of the resultant char material important, though the H/C is to be more accurate than O/C ratio. Compared to other NETs, the co-production of biochar through pyrolysis has been shown to have lower impacts on water use, land, nutrients, energy requirements, and input costs.

Change and Fossil Fuel Offsets Through Land-use

Evaluation of the advantages of using biochar must include the co-production of biofuel as well. The potential yield increase arising from biochar's use could inverse as increased yield decreases the land requirement for food production and energy generation. Kauffman *et al.* find that, assuming a 5.89% increase in corn yield above trend and a 30-year life cycle of biochar, the coproduction of biochar from corn stover can be carbon-neutral. More importantly, with 1% higher yield for corn, soybeans, and wheat in the United States, global land use may decline by 0.06%, or 0.47 million hectares.

Few Important Points

- Agricultural land may be converted for cultivation of crops that feed the coproduction of biofuel and biochar, without damaging food production or forest systems.
- It is obvious that the mitigation potential of the coproduction of biochar and bioenergy depends on the soil fertility and the carbon intensity.
- Biochar helps reduce emissions in the low fertile soils because of the highest improvement in soil productivity and thus in the capacity to capture carbon from the atmosphere.
- Woolf *et al.* ascertain that in all soil-fuel combinations except areas of highest soil fertility, biochar has a higher climate-mitigation potential than bioenergy does though this judgment ignores the fact that liquid fuels for transportation (for aviation industry) are critical to our economy and are not easily replaced with electricity, and that converting biomass into liquid fuels is by far the least expensive way of producing C-free liquid fuels. Allowing this important role of biofuelsit may affect the production choices of bioenergy andbiochar.

Challenges in Realizing Benefits of Biochar

Non-linear interactions in soil-biochar system has led to inconsistent findings for farmers and scientists. This complexity emerges from the dissimilar properties of various biochar products and their interactions with the specific settings of climate, soil, crop, and management. As "Biochar" has its inherent variability controlled by differences in feedstock type, production temperature, and speed [18, 48, 80]. Any alteration in these parameters can yield an extensive range of measured properties including cation exchange capacity, p^H, porosity, surface area, base cation concentration, fixed carbon, etc. Diversity of such properties can produce diverse biochar products that aid altered soil management needs like woody feedstocks can to produce higher fractions of "fixed" or recalcitrant carbon, making them perfect for climate change mitigation, but have relatively lower macronutrient contents and consequently partial efficacy as fertilize. Biochars produced from manures, on the other hand, have substantial levels of nutrients and alkalinity but low fixed carbon, making them superiorliming agents and for recycling nutrients and minor as carbon sequestration agents (Table-3).

Table 3: Variability of biochar from different feedstocks

Sl. no.	Parameter	Biochar from woody feedstock	Biochar from manure
01.	Outcome	woody feedstocks produce higher fractions of recalcitrant (fixed) carbon	Biochars produced from manures have higher concentrations of nutrients and alkalinity but very low fixed carbon
02.	Suitability for climate change	perfect for climate change intervention	less impact as carbon sequestration agents
03.	Suitability for soil	comparatively less macronutrient contents and therefore partial efficacy as fertilizer (Gul & Whalen; Gul et al.; Kloss *et al.*	suitable as liming agents and for recycling nutrients

In view to the complexity, biochar's impact on soil functioning is highly dependent on the soil environment it is applied to, yielding often divergent routes for biochar-soil interactions depending on context. A wide range of analytic methodologies, experimental designs, and target uses have further complicated scientific research on how biochar impacts soil functioning. The complexity of the biochar system and uncertainty about its properties and profitability are challenges that need to be addressed by the supply chain system.

3. Biochar Supply Shain

Biochar supply chain consists of the cost of the main elements of the pyrolysis supply chain and scaling up and supply chain design.

3.1 Cost Driversof the Pyrolysis Supply Shain

The cost of biochar system depends on the following main components:

a) Feedstock acquisition

b) Feedstock transportation

c) Feedstock pretreatment

d) Pyrolysis and operation

Feedstock Acquisition

Several studies find that feedstock sourcing, collection and procurement is one of the most delicate factors in biochar's cost-benefit analysis.

The cost of feedstock sourcing can be estimated as either from

(1) the cost of production if a feedstock is produced specially for pyrolysis, or

(2) the opportunity cost of an existing available feedstock used for pyrolysis.

The cost of sourcing and acquisition for corn stover is merely its market rate [81] whereas used-up grains residues from a brewery can be diverted to bioenergy or to livestock production, thus giving it a relatively higher opportunity cost. On the other hand, if it is a waste stream that otherwise requires a costly treatment (e.g., municipal solid waste) the same may be obtained for a negative price. Thus, with focusing on waste stream feedstocks (e.g. forest thinning, yard waste) may be one of the keys to a profitable pyrolysis unit in the supply chain. The potential synergies of biochar in the forestry context, where feedstock acquisition may incur a negative price. Biochar is most likely to be adopted near locations with marginal land and high-value crop, and near low-cost feedstock sources.

Feedstock Transportation

Transport for feedstock is a linear cost function of distance between the feedstock origin and the location of pyrolysis. Transporting biomass is costly affair as it can add 11% to the production cost. Thus following strategic decisions are important:

- At its very initial stage of technological diffusion biochar production likely be a highly localized market that ensures the location of feedstock supply an important feature.

- Further, as mobile pyrolysers demonstrate their functionality, feedstock proximity may become a less important factor.
- Alternatively, a firm in a complementary industry, like wood and paper products, may opt to co-locate an onsite pyrolysis unit, confirming transport costs to zero.

In the meantime, pyrolysis projects needs to develop nearby the sources offeedstockresources.

Feedstock Pre-treatment

There are limited literatures on pretreatment of feedstock for biochar production system. Pretreatment of feedstock dependson technology available and pyrolysis process as described below:

- slow pyrolysis requires only drying
- drying and coarse chippingis required for gasification with power production
- drying, coarse chipping and then fine grinding (to maximize bio-oilyield, the addition of acid to neutralize the catalytic effects of alkali metals, i.e., "pacification") required for fast pyrolysis with liquid fuels production
- If phosphoric acid is used for pacification purpose, the cost of the acid can be recaptured in the added value of the resulting biochar, which becomes a phosphorous fertilizer product.

Most of the case studies recognize a pretreatment stage (drying and chipping), but do not mention those costs, making it hard to estimate a per-ton average. Campbell, Anderson *et al.* considers a range of feedstock costs, where the lowest cost option is a waste stream that has yet to be chipped, and a high-cost option represents a pre-chipped and dried feedstock. Kung *et al.* assumes pretreatment accounts for about 6% of total production costs across slow and fast pyrolysis processes.One study that can be useful is the relative cost of drying versus chipping. These two processes are largely inversely correlated. Materials from wood need require much drying although require a lot of mechanical energy for chipping. On a contrary, yard waste and sludge involve a sufficient energy input to dry, but little/no chipping is required.

Pyrolysis process operation

The process of actual pyrolysis is responsible for maximum future costs and thus possibility of substantial cost reduction. In the following paragraphs simple and advanced pyrolysis have been discussed individually.

Simple Pyrolysis Process

In simplest methods of pyrolysis involve zero investment input. Pit and mound kilns require little relocation of earth to deliver the preferred effects (e.g., heating with minimal oxygen). Depending upon the extent of capital investments, kilns of several intricacies, shapes and sizes can be made with brick and metal, and can incorporate options such as constant feed arrangements for enhanced performance. If we consider smallholders to manage their own biomass waste with on-farm solutions it is best to adopt simple pyrolysis technology where cost fragments are only the marginal cost for labor for pit making or mounding kilns. Shackley *et al.* haveidentified 10 biochar manufacturing facilities and estimated a per ton cost of pyrolysis. In few slow pyrolysis cases, it is possible to capture process heat it is difficult or no possibility to get bio-oil or associated co-products. To capture these products advanced pyrolysis is most suitable.

Advanced Pyrolysis Process

The viability of fast pyrolysis is most paramount to bio-oil harvest and price, alongwith the price of feedstock, cost of biomass resourcing and other cost input. All calculations are connected to the principal product, i.e., bio-oil, although revenue is relatively less susceptible to the value of biochar as it is considered to be a secondary product here. As compared to gasification and cellulosic ethanol, advanced pyrolysis process is beneficial in terms of the less capital costs and the lowest MFSP (minimum fuel selling price). Consequently, the tradeoff between the production of bio-oil and biochar is suggested to emphasize the choice of different speed and temperature of pyrolysis. According to Brown *et al.*,

- under slow pyrolysis process, the possibility is to get much more biochar and syngas (33%, 66% respectively), and almost no bio-oil (<1%)
- under advanced pyrolysis process returns is over half its energy in bio-oil (53%)

Thus, a remarkable competitive advantage is there when liquid fuels are obtained.

According tothe cost of collection of feedstockas per market-rate slow pyrolysis does not qualify to be a profitable. On a contrary, the study finds that fast pyrolysis may be profitable and benefits momentously in the liquid fuel market in case the gasoline prices are supposed to be increased by 20% between 2015 and 2030, as projected in the proposed American Clean Energy and Security Act (ACESA) (Energy Information Administration, 2009). Both the cost of pyrolysis andsupply chain are significant in the pyrolysis process. McCarl *et al* assessed that syngas and bio-oil can be consumed in energy production. Very close finding by Dickinson *et al.* who find that feedstock collection and transportation contribute about half of the entire production cost, while the another half is ascribed to pyrolysis.

Effect of variation in feedstock

Dissimilarities in feedstock categories and processing technique result in products variations which influence their benefits in different soil environments. Producers of biochar may be benefitted from collaboration as differentiated products of pyrolysis could be isolated at the same site at slight extra cost of diversification by varying feedstock and processing parameters.

3.2 Scaling up and Supply Chain Design

Considering the discussion in the above chapters it can be said that biochar production technology can be treated to be a "premature phase" at its present scenario as the overall economy is yet to establish a steady and viable market. Most of the current works are on cost reduction modules. Full commercialization of biochar necessitates a right design of supply chains for implementation [94], however, biochar supply chains are a part of a larger pyrolysis supply chain.Its full-scale utilization requires

- A better understanding of the chemical and physical properties of different categories of biochar
- Optimal utilization of individual properties
- Design of supply chain on the basis of cost optimization

Biochar supply chains are composed of five essential steps (Table-4, Fig.9):

1. Feedstock selection and acquisition
2. Feedstock transportation
3. Pretreatment of feedstock
4. Pyrolysis process
5. Biochar products application

Table 4: Biochar supply chain

Feedstock Selection& Acquisition	**Feedstock Transportation**	**Pretreatment of Feedstock**	**Pyrolysis Process**	**Products**	**Biochar Applications**	**Benefits of Applications**
• Forests • Agricultural Land (Crops/Residues) • Yard Waste • Manure/Sludge	• Local resource: Low cost • Far from production centre: High cost	• Chipping • Drying • Drying & Chipping	• Simple Pyrolysis • Advanced Pyrolysis	• Biochar • Bio-oil • Syngas	• Agricultural Land • Forests • Remediation Sites	• Improved pH and CEC • Improved nutrient cycling • Improved Water Holding Capacity • Carbon Sequestration

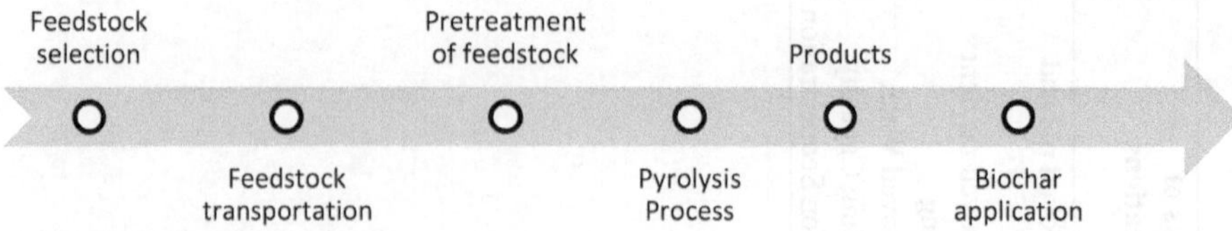

Fig. 9: Supply chain of biochar

The complete pathway of pyrolysis has been reviewed by Campbell, Sessions *et al.*; Campbell, Anderson *et al.*; Galinato *et al.*; Vochozka *et al.*, with consideration of all the elements of supply chain (from feedstock acquisition to deliverables in the energy market). These investigations have recognized an understandable value in generating energy, production of biofuel, management of waste, and sourcing of biochar in the pyrolysis process. Under these studies, a few have emphasized on the strategies of decision making within the framework of supply chain to evaluate the possibility of adoption of biochar as a technology for soil management.

4. Contribution of the study

This paper identifies the individual components of biochar value chain, viz., feedstock selection and acquisition, feedstock transportation, pretreatment of feedstock, pyrolysis process, biochar products application. Considering the initial investment and logistics cost the paper suggests that at the initial stage the biochar unit is supposed to be established nearby the source of biomass. Alternatively, a mobile biochar unit may be designed. In general, it is found that the biochar collection, pretreatment and transportation cost is half of the overall cost of biochar. Thus, it is suggested that if these costs are kept as low as possible, the overall productivity will be sustainable. In this study, the acceptance of biochar technology involves a high carbon price. However, recurrent variations in carbon prices can farmers' investment decisions move away.In the situations where the yield are higher, moderate or even no carbon prices may lead to more adoption.

5. Conclusion and future work: Characterization of biochar

Variations of feedstocks and processing conditions produce biochars with different characteristics. The properties related to surface area, pore volume and diameter, electrical conductivity(EC), cation exchange capacity (CEC), p^H, carbon, hydrogen,H/C, O/C, C/N are important parameters which needs extensive study. The application of such biochars will vary accordingly. The adoption of biochar can be boosted by more extension for carbon sequestration. Planned investment in research, both fundamental and commercial, and education of producers to enhance efficiency of the supply chain, are the paradigms of success of biochar. Characterizing a series of biochar products and their exact capacity to stabilize carbon is critical to comprehend how its production through pyrolysis can mitigate climate change and balance the global carbon fund. A proper analysis of various biochar products and their corresponding stabilities is a prerequisite to the technical inclusion of biochar projects supply chain into carbon sequestration markets.

References

Agegnehu, G., Bass, A. M., Nelson, P. N., Muirhead, B., Wright, G., & Bird, M. I. (2015). Biochar and biochar-compost as soil amendments: Effects on peanut yield, soil properties and greenhouse gas emissions in tropical North Queensland, Australia. Agriculture, Ecosystems & Environment, 213, 72-85.

Agegnehu, G., Srivastava, A. K., & Bird, M. I. (2017). The role of biochar and biochar-compost in improving soil quality and crop performance: A review. Applied Soil Ecology, 119, 156-170.

Al-Wabel, M. I., Al-Omran, A., El-Naggar, A. H., Nadeem, M., & Usman, A. R. (2013). Pyrolysis temperature induced changes in characteristics and chemical composition of biochar produced from conocarpus wastes. Bioresource Technology, 131, 374-379.

Aller, D., Rathke, S., Laird, D., Cruse, R., & Hatfield, J. (2017). Impacts of fresh and aged biochars on plant available water and water use efficiency. Geoderma, 307, 114-121.

Altieri, M. A. (2002). Agroecology: The science of natural resource management for poor farmers in marginal environments. Agriculture, Ecosystems & Environment, 93(1-3), 1-24.

Anderson, N. M., Bergman, R. D., & Page-Dumroese, D. S. (2016). A supply chain approach to biochar systems [Chapter 2]. In Viktor J. Bruckman, Jay Liu, Başak Burcu Uzun, Esin Apaydın-Varol, eds. Biochar: A regional supply chain approach in view of climate change mitigation. Cambridge University Press. p. 25-45.

Anellotech, Inc. (2014). Bio-TCat ™ for renewable chemicals & fuels.

Anex, R. P., Aden, A., Kazi, F. K., Fortman, J., Swanson, R. M., Wright, M. M., Satrio, J. A., Brown, R. C., Daugaard, D. E., & Platon, A. (2010). Techno-economic comparison of biomass-to-transportation fuels via pyrolysis, gasification, and biochemical pathways. Fuel, 89, S29-S35.

Asai, H., Samson, B. K., Stephan, H. M., Songyikhangsuthor, K., Homma, K., Kiyono, Y., Inoue, Y., Shiraiwa, T., & Horie, T. (2009). Biochar amendment techniques for upland rice production in Northern Laos: 1. Soil physical properties, leaf SPAD and grain yield. Field Crops Research, 111(1-2), 81-84.

Bakshi, S., Banik, C., & Laird, D. A. (2020). Estimating the organic oxygen content of biochar. Scientific Reports, 10(1), 1-12.

Baldock, J. A., &Smernik, R. J. (2002). Chemical composition and bioavailability of thermally altered Pinus resinosa(Red pine) wood. Organic Geochemistry, 33(9), 1093-1109.

Bera, T., Collins, H., Alva, A., Purakayastha, T., & Patra, A. (2016). Biochar and manure effluent effects on soil biochemical properties under corn production. Applied Soil Ecology, 107, 360–367.

Biederman, L. A., & Harpole, W. S. (2013). Biochar and its effects on plant productivity and nutrient cycling: a meta-analysis. GCB Bioenergy, 5(2), 202-214.

Brown, T. R., Wright, M. M., & Brown, R. C. (2011). Estimating profitability of two biochar production scenarios: slow pyrolysis vs fast pyrolysis. Biofuels, Bioproducts and Biorefining, 5(1), 54-68.

Bruckman, V. J., Liu, J., & Uzun, B. B. (2016). Biochar. Cambridge University Press.

Campbell, J. L., Sessions, J., Smith, D., & Trippe, K. (2018). Potential carbon storage in biochar made from logging residue: Basic principles and Southern Oregon case studies. PLoS One, 13(9).

Campbell, R. M., Anderson, N. M., Daugaard, D. E., & Naughton, H. T. (2018). Financial viability of biofuel and biochar production from forest biomass in the face of market price volatility and uncertainty. Applied Energy, 230, 330-343.

Cantrell, K. B., Hunt, P. G., Uchimiya, M., Novak, J. M., & Ro, K. S. (2012). Impact of pyrolysis temperature and manure source on physicochemical characteristics of biochar. Bioresource Technology, 107, 419-428.

Cayuela, M., Van Zwieten, L., Singh, B., Jeffery, S., Roig, A., & Sanchez-Monedero, Chen, X., Khanna, M. (2014). Biochar's role in mitigating soil nitrous oxide emissions: A review and meta-analysis.Agriculture, Ecosystems & Environment, 191, 5–16.

Crane-Droesch, A., Abiven, S., Jeffery, S., & Torn, M. S. (2013). Heterogeneous global crop yield response to biochar: A meta-regression analysis. Environmental Research Letters, 8(4), 044049.

Crombie, K., Mašek, O., Sohi, S. P., Brownsort, P., & Cross, A. (2013). The effect of pyrolysis conditions on biochar stability as determined by three methods. GCB Bioenergy, 5(2), 122-131.

Dai, Z., Meng, J., Shi, Q., Xu, B., Lian, Z., Brookes, P., & Xu, J. M. (2016). Effects of manure- and lignocellulose-derived biochars on adsorption and desorption of zinc by acidic types of soil with different properties. European Journal of Soil Science, 67(1), 40-50.

Dickinson, D., Balduccio, L., Buysse, J., Ronsse, F., Van Huylenbroeck, G., & Prins, W. (2015). Cost-benefit analysis of using biochar to improve cereals agriculture. GCB Bioenergy, 7(4), 850-864.

Dokoohaki, H., Miguez, F. E., Laird, D., Horton, R., & Basso, A. S. (2017). Assessing the biochar effects on selected physical properties of a sandy soil: an analytical approach. Communications in Soil Science and Plant Analysis, 48(12), 1387-1398.

Dong, D., Wang, C., Van Zwieten, L., Wang, H., Jiang, P., Zhou, M., & Wu, W. (2020). An effective biochar-based slow-release fertilizer for reducing nitrogen loss in paddy fields. Journal of Soils and Sediments, 20(8), 3027-3040.

Dumortier, J., Dokoohaki, H., Elobeid, A., Hayes, D. J., Laird, D., & Miguez, F. E. (2020). Global land-use and carbon emission implications from biochar application to cropland in the United States. Journal of Cleaner Production, 258, 120684.

Edmonds, J., Luckow, P., Calvin, K., Wise, M., Dooley, J., Kyle, P., Kim, S. H., Patel, P., & Clarke, L. (2013). Can radiative forcing be limited to 2.6 Wm−2 without negative emissions from bioenergy and CO2 capture and storage? Climatic Change, 118(1), 29-43.

El-Naggar, A., Lee, S. S., Rinklebe, J., Farooq, M., Song, H., Sarmah, A. K., Zimmerman, A. R., Ahmad, M., Shaheen, S. M., & Ok, Y. S. (2019). Biochar application to low fertility soils: A review of current status, and future prospects. Geoderma, 337, 536-554.

Fidel, R. B., Laird, D. A., Thompson, M. L., &Lawrinenko, M. (2017). Characterization and quantification of biochar alkalinity. Chemosphere, 167, 367-373.

Field, J. L., Keske, C. M., Birch, G. L., DeFoort, M. W., &Cotrufo, M. F. (2013). Distributed biochar and bioenergy coproduction: a regionally specific case study of environmental benefits and economic impacts. GCB Bioenergy, 5(2), 177-191.

Fuss, S., Canadell, J. G., Peters, G. P., Tavoni, M., Andrew, R. M., Ciais, P., Jackson, R. B., Jones, C. D., Kraxner, F., &Nakicenovic, N. (2014). Betting on negative emissions. Nature Climate Change, 4(10), 850-853.

Galinato, S. P., Yoder, J. K., &Granatstein, D. (2011). The economic value of biochar in crop production and carbon sequestration. Energy Policy, 39(10), 6344-6350.

Gaunt, J. L., & Lehmann, J. (2008). Energy balance and emissions associated with biochar sequestration and pyrolysis bioenergy production. Environmental Science & Technology, 42(11), 4152-4158.

Gul, S., & Whalen, J. K. (2016). Biochemical cycling of nitrogen and phosphorus in biochar-amended soils. Soil Biology and Biochemistry, 103, 1-15.

Gul, S., Whalen, J. K., Thomas, B. W., Sachdeva, V., & Deng, H. (2015). Physico-chemical properties and microbial responses in biochar-amended soils: mechanisms and future directions. Agriculture, Ecosystems & Environment, 206, 46–59.

Hagemann, N., Kammann, C. I., Schmidt, H.-P., Kappler, A., & Behrens, S. (2017). Nitrate capture and slow release in biochar amended compost and soil. PLoS One, 12(2).

Haider, G., Steffens, D., Moser, G., Muller, C., & Kammann, C. I.(2017). Biochar reduced nitrate leaching and improved soil moisture content without yield improvements in a four-year field study. Agriculture, Ecosystems & Environment, 237, 80-94.

Hammes, K., Schmidt, M. W., Smernik, R. J., Currie, L. A., Ball, W. P., Nguyen, T. H., Louchouarn, P., Houel, S., Gustafsson, O., & Elmquist, M. (2007). Comparison of quantification methods to measure fire-derived (black/elemental) carbon in soils and sediments using reference materials from soil, water, sediment and the atmosphere. Global Biogeochemical Cycles, 21(3), 1-18.

Homagain, K., Shahi, C., Luckai, N., & Sharma, M. (2016). Life cycle cost and economic assessment of biochar-based bioenergy production and biochar land application in Northwestern Ontario, Canada. Forest Ecosystems, 3(1), 1-10.

Jeffery, S., Abalos, D., Prodana, M., Bastos, A. C., Van Groenigen, J. W., Hungate, B. A., & Verheijen, F. (2017). Biochar boosts tropical but not temperate crop yields. Environmental Research Letters, 12(5), 053001.

Jin, Z., Chen, C., Chen, X., Hopkins, I., Zhang, X., Han, Z., Jiang, F., & Billy, G. (2019). The crucial factors of soil fertility and rapeseed yield-A five year field trial with biochar addition in upland red soil, China. Science of the Total Environment, 649, 1467-1480.

Kammann, C. I., Schmidt, H.-P., Messerschmidt, N., Linsel, S., Steffens, D., Muller, C., Koyro, H.-W., Conte, P., & Joseph, S. (2015). Plant growth improvement mediated by nitrate capture in co-composted biochar. Scientific Reports, 5(1), 1-13.

Katterer, T., Roobroeck, D., Andren, O., Kimutai, G., Karltun, E., Kirchmann, H., Nyberg, G., Vanlauwe, B., & de Nowina, K. R. (2019). Biochar addition persistently increased soil fertility and yields in maize-soybean rotations over 10 years in sub-humid regions of Kenya. Field Crops Research, 235, 18-26.

Kauffman, N., Dumortier, J., Hayes, D. J., Brown, R. C., & Laird, D. A. (2014). Producing energy while sequestering carbon? The relationship between biochar and agricultural productivity. Biomass and Bioenergy, 63, 167-176.

Keske, C., Godfrey, T., Hoag, D. L., & Abedin, J. (2020). Economic feasibility of biochar and agriculture coproduction from Canadian black spruce forest. Food and Energy Security, 9(1), e188.

Khalifa, N., & Yousef, L. F. (2015). A short report on changes of quality indicators for a sandy textured soil after treatment with biochar produced from fronds of date palm. Energy Procedia, 74, 960-965.

Kim, D., Anderson, N. M., & Chung, W. (2015). Financial performance of a mobile pyrolysis system used to produce biochar from sawmill residues. Forest Products Journal, 65(5-6), 189-197.

Kloss, S., Zehetner, F., Dellantonio, A., Hamid, R., Ottner, F., Liedtke, V., Schwanninger, M., Gerzabek, M. H., & Soja, G. (2012). Characterization of slow pyrolysis biochars: Effects of feedstocks and pyrolysis temperature on biochar properties. Journal of Environmental Quality, 41(4), 990-1000.

Kung, C.-C., McCarl, B. A., & Cao, X. (2013). Economics of pyrolysis-based energy production and biochar utilization: A case study in Taiwan. Energy Policy, 60, 317-323.

Laghari, M., Mirjat, M. S., Hu, Z., Fazal, S., Xiao, B., Hu, M., Chen, Z., & Guo, D. (2015). Effects of biochar application rate on sandy desert soil properties and sorghum growth. Catena, 135, 313-320.

Laird, D., Fleming, P., Wang, B., Horton, R., & Karlen, D. (2010). Biochar impact on nutrient leaching from a Midwestern agricultural soil. Geoderma, 158(3-4), 436-442.

Lashari, M. S., Liu, Y., Li, L., Pan, W., Fu, J., Pan, G., Zheng, J., Zheng, J., Zhang, X., & Yu, X. (2013). Effects of amendment of biochar-manure compost in conjunction with pyroligneous solution on soil quality and wheat yield of a salt-stressed cropland from Central China Great Plain. Field Crops Research,144, 113-118.

Lauber, C. L., Hamady, M., Knight, R., & Fierer, N. (2009). Pyrosequencing-based assessment of soil pH as a predictor of soil bacterial community structure at the continental scale. Applied and Environmental Microbiology, 75(15), 5111-5120.

Lehmann, J., & Joseph, S. (2015). Biochar for environmental management: Science, technology and implementation. Routledge.

Lehmann, J., Gaunt, J., & Rondon, M. (2006). Bio-char sequestration in terrestrial ecosystems-a review. Mitigation and Adaptation Strategies for Global Change, 11(2), 403-427.

Liu, X., Zhang, A., Ji, C., Joseph, S., Bian, R., Li, L., Pan, G., & Paz-Ferreiro, J. (2013). Biochar's effect on crop productivity and the dependence on experimental conditions - A meta-analysis of literature data. Plant and Soil, 373(1), 583-594.

Luo, X., Liu, G., Xia, Y., Chen, L., Jiang, Z., Zheng, H., & Wang, Z. (2017). Use of biochar-compost to improve properties and productivity of the degraded coastal soil in the Yellow River Delta, China. Journal of Soils and Sediments, 17(3), 780-789.

Major, J., Rondon, M., Molina, D., Riha, S. J., & Lehmann, J. (2010). Maize yield and nutrition during 4 years after biochar application to a Colombian savanna oxisol. Plant and Soil, 333(1), 117-128.

Mandal, S., Thangarajan, R., Bolan, N. S., Sarkar, B., Khan, N., Ok, Y. S., & Naidu, R. (2016). Biochar-induced concomitant decrease in ammonia volatilization and increase in nitrogen use efficiency by wheat. Chemosphere, 142, 120-127.

Masiello, C. A., Dugan, B., Brewer, C. E., Spokas, K. A., Novak, J. M., Liu, Z., & Sorrenti, G. (2015). Biochar effects on soil hydrology. In Biochar for environmental management (pp. 575-594). Routledge.

Masson-Delmotte, V., Zhai, P., Portner, H.-O., Roberts, D., Skea, J., Shukla, P. R., Pirani, A., Moufouma-Okia, W., Pean, C., & Pidcock, R. (2018). Global warming of 15°C. Intergovernmental Panel on Climate Change.

McCarl, B. A., Peacocke, C., Chrisman, R., Kung, C.-C., & Sands, R. D. (2012). Economics of biochar production, utilization and greenhouse gas offsets. In Biochar for environmental management (373-390). Routledge.

Mia, S., Van Groenigen, J., Van de Voorde, T., Oram, N., Bezemer, T., Mommer, L., & Jeffery, S. (2014). Biochar application rate affects biological nitrogen fixation in red clover conditional on potassium availability. Agriculture, Ecosystems & Environment, 191, 83-91.

Omondi, M. O., Xia, X., Nahayo, A., Liu, X., Korai, P. K., & Pan, G. (2016). Quantification of biochar effects on soil hydrological properties using meta-analysis of literature data. Geoderma, 274, 28-34.

O'Neill, B., Grossman, J., Tsai, M., Gomes, J., Lehmann, J., Peterson, J., Neves, E., & Thies, J. E. (2009). Bacterial community composition in Brazilian Anthrosols and adjacent soils characterized using culturing and molecular identification. Microbial Ecology, 58(1), 23-35.

Raboin, L.-M., Razafimahafaly, A. H. D., Rabenjarisoa, M. B., Rabary, B., Dusserre, J., & Becquer, T. (2016). Improving the fertility of tropical acid soils: Liming versus biochar application? A long term comparison in the highlands of Madagascar. Field Crops Research, 199, 99-108.

Randolph, P., Bansode, R., Hassan, O., Rehrah, D., Ravella, R., Reddy, M., Watts, D., Novak, J., &Ahmedna, M. (2017). Effect of biochars produced from solid organic municipal waste on soil quality parameters. Journal of Environmental Management, 192, 271-280.

Rawls, W., Pachepsky, Y. A., Ritchie, J., Sobecki, T., & Bloodworth, H. (2003). Effect of soil organic carbon on soil water retention. Geoderma, 116(1-2), 61-76.

Riahi, K., Kriegler, E., Johnson, N., Bertram, C., Den Elzen, M., Eom, J., Schaeffer, M., Edmonds, J., Isaac, M., & Krey, V. (2015). Locked into Copenhagen pledges-implications of short-term emission targets for the cost and feasibility of long-term climate goals. Technological Forecasting and Social Change, 90, 8-23.

Roberts, K. G., Gloy, B. A., Joseph, S., Scott, N. R., & Lehmann, J. (2010). Life cycle assessment of biochar systems: Estimating the energetic, economic, and climate change potential. Environmental Science & Technology, 44(2), 827-833.

Rogelj, J., Luderer, G., Pietzcker, R. C., Kriegler, E., Schaeffer, M., Krey, V., & Riahi, K. (2015). Energy system transformations for limiting end-of- century warming to below 1.5°C. Nature Climate Change, 5(6), 519-527.

Sha, Z., Li, Q., Lv, T., Misselbrook, T., & Liu, X. (2019). Response of ammonia volatilization to biochar addition: A meta-analysis. Science of the Total Environment, 655, 1387-1396.

Shackley, S., Clare, A., Joseph, S., McCarl, B. A., & Schmidt, H.-P. (2015). Economic evaluation of biochar systems: Current evidence and challenges. In Biochar for environmental management (845-884). Routledge.

Shepherd, J. G., Buss, W., Sohi, S. P., & Heal, K. V. (2017). Bioavailability of phosphorus, other nutrients and potentially toxic elements from marginal biomass-derived biochar assessed in barley (Hordeum vulgare) growth experiments. Science of the Total Environment, 584, 448-457.

Smith, P. (2016). Soil carbon sequestration and biochar as negative emission technologies. Global Change Biology, 22(3), 1315-1324.

Sohi, S. P., Krull, E., Lopez-Capel, E., & Bol, R. (2010). A review of biochar and its use and function in soil. Advances in Agronomy, 105, 47-82.

Soja, G., Anders, E., Bucker, J., Feichtmair, S., Gunczy, S., Karer, J., Kitzler, B., Klinglmuller, M., Kloss, S., & Lauer, M. (2016). Biochar applications to agricultural soils in temperate climates–more than carbon sequestration. In Biochar: A regional supply chain approach in view of climate change mitigation (pp. 291-314). Cambridge University Press.

Spokas, K. A. (2010). Review of the stability of biochar in soils: Predictability of O:C molar ratios. Carbon Management, 1(2), 289-303.

Steiner, C., Das, K., Melear, N., &Lakly, D. (2010). Reducing nitrogen loss during poultry litter composting using biochar. Journal of Environmental Quality, 39(4), 1236-1242.

Suliman, W., Harsh, J. B., Abu-Lail, N. I., Fortuna, A.-M., Dallmeyer, I., & Garcia-Perez, M. (2016). Influence of feedstock source and pyrolysis temperature on biochar bulk and surface properties. Biomass and Bioenergy, 84, 37-48.

Thompson, J. L., & Tyner, W. E. (2014). Corn stover for bioenergy production: Cost estimates and farmer supply response. Biomass and Bioenergy, 62, 166-173.

Tisserant, A., & Cherubini, F. (2019). Potentials, limitations, co- benefits, and trade-offs of biochar applications to soils for climate change mitigation. Land, 8(12), 179.

Van Zwieten, L., Kimber, S., Downie, A., Morris, S., Petty, S., Rust, J., & Chan, K. (2010). A glasshouse study on the interaction of low mineral ash biochar with nitrogen in a sandy soil. Soil Research, 48(7), 569-576.

Vispute, T. P., Zhang, H., Sanna, A., Xiao, R., & Huber, G. W. (2010). Renewable chemical commodity feedstocks from integrated catalytic processing of pyrolysis oils. Science, 330(6008), 1222-1227.

Vochozka, M., Maroušková, A., Vachal, J., &Strakova, J. (2016). Biochar pricing hampers biochar farming. Clean Technologies and Environmental Policy, 18(4), 1225-1231.

Wang, D., Li, C., Parikh, S. J., & Scow, K. M. (2019). Impact of biochar on water retention of two agricultural soils-A multi-scale analysis. Geoderma, 340, 185-191.

Weil, R. R., & Brady, N. C. (2016). The nature and properties of soils (15th ed.). Pearson.

Woolf, D., Amonette, J. E., Street-Perrott, F. A., Lehmann, J., & Joseph, S. (2010). Sustainable biochar to mitigate global climate change. Nature Communications, 1(1), 1-9.

Yuan, J.-H., Xu, R.-K., & Zhang, H. (2011). The forms of alkalis in the biochar produced from crop residues at different temperatures. Bioresource Technology, 102(3), 3488-3497.

Zhang, A., Liu, Y., Pan, G., Hussain, Q., Li, L., Zheng, J., & Zhang, X. (2012). Effect of biochar amendment on maize yield and greenhouse gas emissions from a soil organic carbon poor calcareous loamy soil from Central China Plain. Plant and Soil, 351(1), 263-275.

Zhang, L., & Sun, X. (2014). Changes in physical, chemical, and microbiological properties during the two-stage co-composting of green waste with spent mushroom compost and biochar. Bioresource Technology, 171, 274-284.

Zhang, R., Zhang, Y., Song, L., Song, X., Hanninen, H., & Wu, J. (2017). Biochar enhances nut quality of Torreya grandis and soil fertility under simulated nitrogen deposition. Forest Ecology and Management, 391, 321-329.

Zheng, J., Han, J., Liu, Z., Xia, W., Zhang, X., Li, L., Liu, X., Bian, R., Cheng, K., & Zheng, J. (2017). Biochar compound fertilizer increases nitrogen productivity and economic benefits but decreases carbon emission of maize production. Agriculture, Ecosystems & Environment, 241, 70-78.

Zilberman, D., Lu, L., & Reardon, T. (2019). Innovation-induced food supply chain design. Food Policy, 83, 289-297.

Zilberman, D., Zhao, J., & Heiman, A. (2012). Adoption versus adaptation, with emphasis on climate change. Annual Review of Resource Economics, 4(1), 27-53.

12

Mitigation of Human-Wildlife Conflict to Achieve Sustainable Development Goals

Samala Akhila*, Bharat S Sontakki

**Professor Jayashankar Telangana State Agricultural University Rajendranagar, Hyderabad, Telangana*
Extension Systems Management Division, ICAR-National Academy of Agricultural Research Management, Hyderabad, Telangana

Abstract

Agriculture is one of the most strategic sectors to eradicate extreme poverty, increase shared prosperity, and feed an estimated 9.7 billion people by 2050. Many strategic options are being explored to meet the food, fibre and nutritional requirements for ever increasing population. A significant area is being converted to agriculture in order to meet the demands of the expanding population. This conversion involves switching from naturally occurring ecosystems to human-managed ones, which shatters wild animal habitats and forces them into cropland habitats where they severely harm standing crops. Goals 1 (No Poverty), 2 (Zero Hunger) and 15 (Life on Land) are very pertinent to reducing Human Wildlife Conflict; hence actions made by the national and state governments, as well as by all other institutions, would directly support the achievement of these three SDGs for India. To attain the SDGs, more emphasis must be paid to the human-wildlife conflict, which affects both development and humanitarian issues in addition to conservation concerns. Achieving the UN Vision for Biodiversity 2050, which calls for "people to live in harmony with environment and in which wildlife and other living species are safeguarded," would consequently require managing human-animal conflicts. Such human-wildlife conflict management solutions may generate possibilities and advantages not just for biodiversity and impacted people, but also for society, sustainable development, production, and the global economy as a whole.

Keywords: Human-Wildlife Conflict, Sustainable Development Goals, Biodiversity

Introduction

Agriculture has historically been the primary source of income in India. It has the potential to alleviate poverty, increase income, and enhances food security for 80% of both the world's poor and people who reside in rural areas primarily work in agriculture. One of the most effective strategies to eradicate extreme poverty, increase shared prosperity, and feed an estimated 9.7 billion people by 2050 is agricultural development. Compared to other modes, the agro - based sector's progress is two to four times more successful in enhancing the earnings of the poorest people (www.worldbank.org).

Nearly 5 billion hectares, or 38% of the world's land surface, is used for agriculture. Around one of this is devoted for farming, and the remaining two-thirds are pastures and meadows for grazing animals. Over 10% of cropland is assigned to permanent crops like cocoa plantations, oil palm plantations, and fruit tree orchards. Yet another 21% of land is primed for irrigation, a crucial agricultural land management technique.

Agriculture is indeed vital to the economy, taking account for 4% of global GDP and more than 25% of GDP in some least developed countries (www.worldbank.org). In recent decades, increases in crop yields have been noted in order to meet the rising food demand of a growing global population. However, advancement has frequently been accompanied by social and environmental costs, such as water scarcity, soil erosion, ecosystem stress, biodiversity loss, declining fish stocks and forest cover, and high greenhouse gas emissions.

Today, Food and agriculture are at a crossroads. Crop yields could be even more reduced by expediting climate change, particularly in areas with the greatest food insecurity in the world. Approximately 25% of greenhouse gas emissions are attributable to agriculture, forestry, and changes in land use. However, food security, poverty reduction, and agricultural based growth are all in jeopardy: Numerous shocks, including COVID-19-related breakdowns, Biodiversity loss severe weather, pests, and conflicts, are having an impact on food systems, driving up food prices and escalating hunger.

As the World population is burgeoning gradually the demand for food is also being increased. A significant area is being converted to agriculture in order to meet the demands of the expanding population. This conversion involves switching from naturally occurring ecosystems to human-managed ones, which shatters wild animal habitats and forces them into cropland habitats where they severely harm standing crops.

An unbalanced food system can be seen in the fact that 820 million people are undernourished and one in three is hungry today (www.un.org). Anguish relocation is at levels that haven't been seen in more than 70 years as rural populations' social cohesiveness and cultural traditions have been confronted by a convergence of restricted availability of resources and land, an increase in crises, conflicts, and disasters, many of which are a result of climate change.

The 2030 Agenda for Sustainable Development evidently outlines the way to inclusive prosperity in the future. Food and agriculture, which serve as the primary link among both people and the environment, can aid in achieving a number of the Sustainable Development Goals (SDGs).

The United Nations adopted a set of targets in 2015 that envisions a world free from poverty and hunger in 2030 as well as protection from the worst effects of climate change and biodiversity loss.

SDG Goal 1(No Poverty), Goal 2 (Zero Hunger) and 15 (Life on Land) are very pertinent to reducing Human Wildlife Conflict; hence actions made by the national and state governments, as well as by all other institutions, would directly support the achievement of these two SDGs for India. To attain the SDGs, more emphasis must be paid to the human-wildlife conflict, which affects both development and humanitarian issues in addition to conservation concerns.

What is Human-Wildlife Conflict?

According to World Wildlife Fund Human-Wildlife Conflict refers to any interaction between humans and wildlife those results in the negative impacts of human social, economic or cultural life on the conservation of wildlife populations, or the government.

Why it is Important?

According to World Wildlife Day 2020, human-wildlife conflict in India affects about 90% of the country's protected areas, 88 species, and 50 million people.

Therefore, those living in rural areas that rely heavily on ecosystem services for their livelihoods and who are frequently least able to acquire or purchase alternatives for the lost ecosystem services suffer the severe effects of biodiversity loss and environmental degradation. As human populations increase and ecosystems are destroyed, confrontations between humans and wildlife are getting more frequent, serious, and pervasive.

Many efforts were made up to date to prevent negative effects on people and reprisal against animals by erecting barriers, using deterrents, or transporting wildlife. These measures frequently have poor results in the absence of deliberative, collaborative approaches involving stakeholders.

Long-term advantages of ecosystem, biodiversity, and animal conservation outweigh any trade-offs that may arise from the actions of different sectors.

Achieving the UN Vision for Biodiversity 2050, which calls for "people to live in harmony with environment and in which wildlife and other living species are safeguarded," would consequently require managing human-animal conflicts (www.iucn.org).

Causes of Human Wildlife Conflict

Human-animal conflicts arise in regions where human and wildlife populations coexist, even on the edges of protected areas, and it is harmful to both species. Conflict between humans and wildlife has a variety of location factors.

Conflict between humans and animals affects a wide range of species and occurs in a number of contexts, having direct and indirect, material and intangible effects.

Population Growth: As human populations increase, settlements move into and through protected areas, causing encroachment into animal habitats. These demographic and socioeconomic changes bring more humans into direct contact with wildlife.

Land use Change: As a result of the rising need for land, food production, energy, and raw resources, forests and other ecosystems are being converted into agricultural regions or urban areas.

Climate Change: Because of increased resource scarcity and increased pressure to coexist in congested areas, climate change is escalating human-wildlife conflicts are being made worse by climate change, which is stressing ecosystems and changing behaviour, both of which might increase human-animal interactions and perhaps increase competition.

Abundance of Prey: When natural food is ample, carnivorous animals prefer it over livestock, and prey population depletion is one of the primary drivers of carnivores changing their diets to livestock.

Species habitat loss, degradation: Because of Expansion of Agricultural land in order to meet the demands of growing population the forest areas are being reduced. As a result of biodiversity loss, humans and wildlife are competing for limited resources in shared landscapes, which often has unintended consequences.

Strategies to Reduce Human-Wildlife Conflict?

1. HWC prevention

Itrefers to an Effective HWC management i.e., to take steps to prevent or reduce HWC before it happens. Crop selection, early warning systems, fence and repellent application are a few HWC prevention strategies (A Future for All: The Need for Human-Wildlife Coexistence, 2021).

Crop Selection: Choosing crops that will not be damaged by wildlife, diversifying into different types of crops, using an intercropping scheme, and altering the date of harvest may all assist to reduce conflict.

Early Warning Systems: Early warning systems enable humans to take preventive action by seeing animals before it might cause harm. The trip-wire system is the most basic early warning system used to find wild pigs and elephants. This is a straightforward wire that is stretched throughout the perimeter of the agricultural field and connected to a noisemaker, typically a firecracker.

Fences and Repellent Application: The main purpose of barriers like Barbed wire, Mesh wire and Electric fences etc. is to keep animals out of locations where people congregate. The ideal location for barriers is where human habitation and wildlife habitat converge. Utilizing a combination of natural repellents, such as predator urine or chemical repellent, around the field lessens the original smell of the crop, which prevents wild animals from entering fields by masking the smell.

2. Habitat Management: The loss of resources available to animals due to habitat fragmentation is frequently a primary factor in HWC. Prior to fragmentation, there is frequently communication between forest sections via corridors. Over time, these corridors are vulnerable to breaking down for a variety of reasons, including habitat loss or degradation, an increase in disturbances, the edge effect etc.

a) Extension and Communication Strategies

- A vital part of managing and conserving biodiversity is raising awareness of HWC and starting conversations about it. In order for them to effectively engage in such procedures, it is crucial that key stakeholders be informed of the biodiversity management efforts and HWC mitigation strategies.
- Activities for education and training at various levels, such as those conducted in classrooms or adult education settings like farmer field

schools, would aim to spread cutting-edge methods, strengthen community capacity for conflict resolution, and broaden public awareness of HWC.

- Primary Response teams (PRT), Rapid Response Teams (RRT) and Reporting Systems must be established in all districts and they should be trained in order to combat with Human-Wildlife Conflict.
- Gathering and examining data on HWC and potential root causes, including local and scientific knowledge.
- Extension agents are the important change agents that they will change the negative perception and attitude of rural people because of Conflict.
- Rural people would benefit from receiving practical training in order to deal with dangerous wild animal species and to acquire and create new weapons for defending their livestock and crops. As a consequence, local populations would gradually alter their behaviour, which would lower hazards, enhance local livelihoods, and lessen locals people Vulnerability.

Human- Wildlife Conflict Mitigation

It refers to any action taken to lessen the effects of HWC after they have already happened, such as the creation of alternative livelihoods, insurance plans, or compensation programmes etc. (A Future for All: The Need For Human-Wildlife Coexistence, 2021)

- Skill development trainings and Education awareness programmes should be provided to youth that will enable to search for alternate livelihoods options.
- Community Camera trapping Program to strengthen connections between community advantages and the presence of wildlife in the area (FAO and IUCN SSC HWCTF. 2022).
- **Compensation Programmes:** The Compensation which is referred as Ex-Gratia is paid in different amounts according to the rules of a particular state.

Example: Maharashtra pays ex-gratia for the damage caused by Herbivores while some states pay Ex-gratia if the crop is damaged by elephants.

Conclusion

As our world gets more congested because of Increase in human Population, Human-Wildlife Conflict will always persist; nevertheless, competent, well-planned management and comprehensive and integrated methods can lessen and limit conflict in the long run. Such human-wildlife conflict management solutions may generate possibilities and advantages not just for biodiversity and impacted people, but also for society, sustainable development, production, and the global economy as a whole. We also want worldwide collaboration, coordinated measures, and resources to solve the issue at the necessary level.

To combat HWC, no single technique is sufficient. Clear strategies for preventing and dealing with HWC may aid in the development of solutions that can be adopted Administrations (national or municipal), wildlife agencies, farmers and communities, and/or the commercial sector are all involved. Policies should be developed collaboratively by all stakeholders, particularly local communities.

Extension personnel play a key role in mitigating Human-Wildlife Conflict because they are the front line workers who have direct contact with the farmers to provide training programmes to mitigate the conflict and to change the perception, Attitude of rural people and ensure living with Harmony in nature.

Strategies to prevent, minimise, and manage HWC will be standardised throughout all states in the nation with regard to the different types of response teams, their purpose, duties, makeup, and necessary competences.

The ability of managers to acknowledge, accept, and take into account various stakeholder values, attitudes, and beliefs is crucial to the effectiveness of wildlife conservation and HWC reduction.

A HWC Mitigation Hub/HWC Control Room will be used to coordinate the management of information across teams. As competence and specialised knowledge are required for the job, such RRTs will maintain a steady pool of trained employees. Options for enlisting specialisedorganisations and individuals, such as Extension personnel, civil defence, house guards, and disaster management teams, to assist the reaction teams. As and when possible, the involvement of local body officials, such as gram panchayats, affected community can also be solicited in order to improve coordination at the local level (HWC-NAP 2021).

References

Bhavani, R.V. and Rampal, P. (2018). Review of agriculture – nutrition linkages inSouth Asia. Retrieved from http://www.cabi.org/cabreviews (accessed on December 27, 2022).

FAO and IUCN SSC HWCTF. 2022. Co-developing a community camera trapping programme to deliver benefits of living with wildlife. Rome, FAO.

Gross, E., Jayasinghe, N., Brooks, A., Polet, G., Wadhwa, R. and Hilderink-Koopmans, F. 2021. A Future for All: The Need for Human-Wildlife Coexistence. (WWF,Gland, Switzerland).

https://www.iucn.org/resources/issues-brief/human-wildlife-conflict

https://www.worldbank.org/en/topic/agriculture/overview

Ministry of Environment, Forest and Climate Change, Government of India. 2021. National Human-Wildlife Conflict Mitigation Strategy and Action Plan of India-2021-26.

UNEP, 2021. Human Wildlife Conflict one of the greatest threats to wildlife species- WWF and UNEP report.

13

Climate Smart Village An Approach for Sustainable Development

Shivendra Mishra, Mani Prakash Shukla and Pankaj Kumar Ojha

Department of Agricultural Extension, Banda University of Agriculture and Technology, Banda, Uttar Pradesh

Abstract

Over the past few decades, climate change has adversely affected both physical and biological systems in most of the countries across the globe, which leads to decrease in agricultural production by 1-5% per decade. To mitigate these changes a new concept that is climate smart village come in existence. The term Climate Smart Village (CSV) has become more popular to mitigate climate change and its adverse effects at grass root level. The objective of Climate Smart Village (CSV) model is ensuring food security, increasing adaptive capacity and mitigating adverse effect of climate change. A CSV should be weather smart, water smart, carbon smart, nitrogen smart, energy smart and knowledge smart. Recent year's agricultural system needs some climate resilient and sustainable agricultural practices, so the Climate Smart Village (CSV) is a mean of sustainable development at grass root level and it is also a demand of future to mitigate the adverse effects of climate change.

Keywords: CSV, Mitigation, Participation, Climate resilient agriculture, Sustainable development

Introduction

Despite recent significant increases in global food production Sustainable development goal- "End hunger, achieve food security and improved nutrition, and promote sustainable agriculture" seems unachievable. Globally, 1 in 9 people are undernourished, the vast majority of whom live in developing countries. Under nutrition causes wasting or severe wasting of 52 million children worldwide. It contributes to nearly half (45%) of deaths in children under five – 3.1 million children per.Today, the majority of the world's hungry people—nearly 870 million—live in sub-Saharan Africa and South Asia. Climate change impacts on food security and livelihoods are already alarming and affecting millions of smallholder farmers in sub-Saharan Africa (FAO, 2016).

In view of today's 795 million hungry and the additional 2 billion people expected by 2050, there is an urgent need for a change of global food and agriculture systems in order to sustainably increase food production (UN, 2016). This will exert intense pressure on agro-ecosystems that are already overburdened, particularly in developing countries as most of those increases in production have to occur in this part of the world. The environmental impacts of satisfying this increased food demand will be further aggravated by climate change. With increased frequency and severity of droughts, extreme heat conditions and over dependence on rainfed agriculture, there is a growing agricultural productivity crisis, dwindling household food availability and the economic prosperity of countries whose national economies are dependent on agriculture. Considering that climate change impacts are felt differently within regions, context-specific adaptation measures are required to reduce risks and build adaptive capacity of smallholder farmers (Altieri, M.A*et al.*2017).

Thus, it is expected to impact more in already-vulnerable developing regions with relatively lower technical and economic capacity to respond to these threats. So, to address this problem of climate change and to increase production in a sustainable manner CGIAR introduced a Research Program on Climate Change, Agriculture and Food Security (CCAFS) in 2009 in different countries of the world especially the more prone areas to climate change. Under the same projects Climate-Smart Villages (CSVs) were established in Southeast Asia to serve as models of climate-resilient communities and field laboratories of climate-smart agriculture (CSA).The CSV approach is a multispectral platform for testing technological and institutional options for climate change adaptation and mitigation in agriculture. Highlighting context, processes, and outcomes, the CSV concept is proven to be an innovative approach to mainstream CSA in rural agrarian communities and surrounding landscapes.

Climate Smart Agriculture (CSA)

CSA refers to packages of technologies/agricultural practices that addresses food security and climate challenges by sustainably increasing agricultural productivity and income, adapting and building resilience to climate change, and reducing and/or removing GHG emission, where possible (Dinesh *et al.*, 2015; FAO, 2012, 2013).

Many organisations have been working on CSA for last two decadesand focusing more on the technology aspects of CSA. However, the institutional and governance aspects at local level need to be well integrated into it while these are the most crucial issues according to the review of CSA by (Dinesh *et al.* 2015).

Three Pillars of Climate Smart Agriculture

These pillars also describe the objectives of Climate Smart Village (CSV). According to Papuso and JimlyFaraby (2013) these are the three pillars of CSA-

- **Productivity:** CSA aims to sustainably increase agricultural productivity and incomes from crops, livestock and fish, without having a negative impact on the environment. This, in turn, will raise food and nutritional security. A key concept related to raising productivity is sustainable intensification.
- **Adaptation:** CSA aims to reduce the exposure of farmers to short-term risks, while also strengthening their resilience by building their capacity to adapt and prosper in the face of shocks and longer-term stresses. Particular attention is given to protecting the ecosystem services which ecosystems provide to farmers and others. These services are essential for maintaining productivity and our ability to adapt to climate changes.
- **Mitigation:** Wherever and whenever possible, CSA should help to reduce and/or remove greenhouse gas (GHG) emissions. This implies that we reduce emissions for each calorie or kilo of food, fibre and fuel that we produce. That we avoid deforestation from agriculture. And that we manage soils and trees in ways that maximizes their potential to acts as carbon sinks and absorb CO2 from the atmosphere.

Climate-Smart Villages

A CSV is a site where researchers from national and international organisations, members of farmers' cooperatives, local government, community leaders, private sector organisations and key policy planners come together to identify the climate-smart agriculture interventions that are most appropriate to tackle the climate challenges of the agriculture sector in the village (Aryal *et al.*, 2015).

Climate-smart village (CSV) is an institutional approach to test, implement, modify and promote CSA at the local level and to enhance farmers' abilities to adapt to climate change, manage climate risks and build resilience. The

CSV adopts a portfolio of interventions that cover the full spectrum of farm activities. The local-level unit can be defined in terms of community or village or cluster of villages. In south Asia, CCAFS considers 'village' as a local unit to implement its CSA and therefore, a village where one or several of the CSAs are tested/implemented, is called a CSV.

CCAFS (climate change agriculture and food security) developed and implemented the CSV model to generate evidence on the effectiveness of climate-smart agriculture. Being implemented in South and Southeast Asia, East and West Africa, and Latin America, the CSV model aims "to generate evidence at local scales of what CSA options work best, where, why, and how, and use this evidence to draw out lessons for policymakers, agricultural development practitioners, and investors from local to global levels" (Aggarwal et al. 2018).

Specifically, the CSV model aims to: evaluate CSA options for productivity, adaptation, and mitigation; develop solutions for future climate change impacts; learn the factors affecting adoption; and identify effective socially inclusive and integrative financial and policy tools (Aggarwal et al, 2018). Major components of a typical CSV model are climate-smart agriculture technologies and practices (T&Ps), climate information services, insurance, institutions, policies, and finance (Aggarwal et al, 2018).

Origin of CSV

The CGIAR (Consultative group for International Agricultural Research) is the world's largest research partnership for greater food security and sustainable agriculture worldwide. It focuses its science and partnerships on helping farmers and their communities become more resilient in the face of countless challenges and threats. CGIAR, established a Research Programme on CCAFS (Climate Change Agriculture and Food Security) in 2009 as a 'Challenge Program' (later renamed a 'Research Program') to work in a collaborative way across the organisationto overcome the threats caused by a changing climate and to achieving food security, enhancing livelihoods and improving environmental management in the developing world. CCAFS brings together the world's best strategic research in agricultural sciences, development research, climate science, and earth system science to identify and address the most important interactions, synergies and trade-offs between climate change, agriculture in a broad sense and food security. As a collective effort, CCAFS has become a hub that facilitates action across multiple CGIAR centers and other CGIAR Research Programmes while involving farmers, development agents, policy makers, donors and other stakeholders to integrate their knowledge and needs into the tools and approaches that are developed.

Key interventions in a Climate-Smart Village

According to Aggarwal *et al*(2013) the different practices followed in a CSV which makes it smart in terms of weather, water, knowledge and so on, are as follows:

Weather Smart: Information related to weather in is very important in Climate Smart Villages (CSV). Farmers are linked to weather information and value-added agro-advisories through radio shows, television, newspapers and mobile phone voice messages. Farmers can use indexbased insurance schemes to cover risks associated with changes in rainfall and temperature at the different stages of crop growth. Timely and appropriate information on crop management particularly those related application of inputs like fertiliser, pesticides and so forth are crucial for farmers to makethe right management decisions. For example, a farmer in Uttar Pradesh reported that he was advised to spray certain types of pesticides (Quinalphos 25EC @ 3ml/litre of water)for controlling pests in mustard. This helped him to earn approximately USD1000 per acre additional income. (http://www.iksl.in/).

Water Smart: Resilient water management practices which aim at enhancing the efficiency and productivity of water are critical climatesmart interventions. These could include aquifer recharge, rainwater harvesting, and community management of water, laser land leveling, water conservation, drip irrigation and onfarm water management practices. For instance, drainage is important in places that are prone to becoming waterlogged, such as Bihar in India and Bangladesh. A study by Aryal et al., (2015) shows that crop yields are higher in laser levelled fields compared to traditionally-levelled ones, whereas the total duration of irrigation per ha per season was much lower in laser-levelled fields. This corresponds to water-savings and to energy saving due to reduced fuel requirements forthe operation of tube wells for irrigation.

Carbon Smart: Soil plays an important role in the global carbon cycle. The soil carbon pool has beenestimated at approximately 3.3 times the size of the atmospheric pool (Lal, 2004).Carbon content in the soil can be increased through agricultural practices such asagroforestry, livestock and manure management, conservation tillage, diversified land-use systems and residue management.

Nitrogen Smart: In Climate-Smart Villages, farmers use leaf-color charts, handheld crop sensors, and nutrient decision-maker tools to decide the most appropriate dosage of nitrogen fertilizers for their crops. This saves on costs and also cuts down on greenhouse gas (GHG) emissions.Recent approaches like site-specific nutrient management (SSNM) provide an approach to 'feeding crops' with nutrients as and when they are needed and hence making synergy for nutrient demand and supply under a certain production system, thereby increasing the nutrient-use efficiency and minimising the nutrient losses into the environment (Jat et al., 2013; Sapkota et al., 2014).

Energy Smart: The energy sector, which produces nearly 60% of CO2 emissions, is the largest contributor to climate change (FAO, 2011). Globally, energy demand is increasing and so is the competition between agriculture and the non agricultural sectors for energy use (FAO, 2013). Therefore, improving energy use efficiency of agricultural production is an effective way to reduce production costs and lower emissions.CCAFS promotes fuel-efficient agro machineries, residue management and reduced tillage as interventions to conserve energy and reduce GHG emissions. In some cases, biogas systems are promoted using manure slurry from intensive dairy enterprises as part of the portfolio of innovations.

Knowledge Smart:CCAFS and partner organizations arrange cross-site visits of farmers to analogue sites and to other areas practicing climatesmart agriculture. An innovative approach of crowdsourcing seeds is used in Climate-Smart Villages. A large number of farmers are provided with seed packets of adapted varieties to evaluate those best suited to their local conditions. They provide feedback to researchers to help them develop better varieties.

Setting up a Climate-Smart Village

There are seven steps to setting up a climate smart village. The climate smart villages are selected after the base survey of Village Climate Risk Management Committee (VCRMC)(Gonsalves, 2020). The different steps are given as follows:

1. **Selecting the site:** The location of a Climate-Smart Village is selected based on its climate risk profile, alternate land-use options, and on the willingness of farmers and local government to participate in the project.
2. **Working with communities**: Community involvement is integral to the success of a Climate-Smart Village. CCAFS forms, or works with existing community groups, consisting of farmers, researchers, rural agro-advisory service providers and village officials. They are briefed on the objectives of Climate-Smart Villages and encouraged to formally register with the government (if they have not already) to benefit from subsidies on government schemes.

3. **Conducting the baseline survey:** Researchers conduct a comprehensive baseline study to capture the current socio-economic situation, resource availability, average production and income and risk management approaches of village households. This enables an assessment of the impact of the interventions after a certain period of time.

4. **Prioritizing interventions:** Stakeholders convene to prioritize and test which climate-smart technologies and approaches are best suited to their local conditions. Focus group discussions involve farmers in a choice experiment using dummy money to indicate which actions they would most willingly carry out.

5. **Building capacity:** To promote the community's involvement and motivate farmers, a range of tools and approaches are sometimes offered up front. These include rain gauges, improved seed varieties, new livestock breeds, tree seedlings, simple machinery such as zero-till machines, subsidies on index-based insurance premiums and discounts on cell phone SIM cards. Scientists, private sector representatives, and local government organize regular training sessions for farmers on good agricultural practices. At some sites a small farm is used by the researchers to demonstrate the complete portfolio of interventions.

6. **Monitoring and evaluating progress:** The lead partner in the village appoints a site coordinator and assistant to provide technical inputs and liaise with CCAFS resource persons. Participating farmers maintain a daily diary of their farm activities and work with the site coordinator to monitor and evaluate the progress of their chosen interventions. These results are digitized and analyzed by researchers at the end of every crop season.

7. **Disseminating outcomes:** To spread the message of climate-smart agriculture, participatory videos on success stories and testimonials from the pilot villages are screened in nearby villages. Success stories are also widely publicized through local, national and international media. Local government partners organize regular 'farmer field days' to motivate farmers, address their questions and improve on existing strategies.

Conclusion

Climate change has really become a big challenge for whole world, especially for developing country. It's adverse effects, affects the agricultural production system and food, health and nutritional security. Many policies makes and researchers give emphasis on developing climate smart technology, climate smart farming methods for mitigate itsadverse effect for sustainable development. The climate smart village may be very effective model for mitigate climate change at grass root level. It takes care of all the aspects of livelihood security at village level in changing climate scenario. It provides guideline for climate smart activity related to agriculture and livestock production as well as sustainable resources development and consumption.

References

Aggarwal PK, Jarvis A, Campbell BM, Zougmoré, RB, Khatri-Chhetri A, Vermeulen SJ, Loboguerrero A, Sebastian L S, Kinyangi J, Bonilla-Findji O, Radeny M, Recha J, Martinez-Baron D, Ramirez-Villegas J, Huyer S, Thornton P, Wollenberg E, Hansen J, Alvarez-Toro P, Aguilar-Ariza A, Arango-Londoño D, Patiño-Bravo V, Rivera O, Ouedraogo M, BTY. 2018. The climate-smart village approach: framework of an integrative strategy for scaling up adaptation options in agriculture. Ecology and Society 23(1):14. https://doi.org/10.5751/ES-09844-230114.

Altieri, M.A.; Nicholls, C.I. The adaptation and mitigation potential of traditional agriculture in a changing climate. Clim. Chang. 2017, 140, 33–45. [Google Scholar] [CrossRef]

Aryal, J.P., Jat, M., Singh, R., Gehlawat, S.K. and Agarwal, T. (2015) Framework, Guidelines and Governance for Designing Local Adaptation Plan of Action to Mainstream Climate-Smart Villages in India, Internantional Maize and Wheat Improvement Center (CIMMYT), New Delhi, India.

Dinesh, D., Frid-Nielsen, S., Norman, J., Mutamba, M., Loboguerrero Rodriguez, A. And Campbell, B. (2015) Is Climate-Smart Agriculture Effectivc? A Review of Selected Cases. CCAFS Working Paper No. 129, Copenhagen, Denmark: CGIAR Research Program on Climate Change, Agriculture and Food Security (CCAFS), Retrieved fromhttps://cgspace.cgiar.org/bitstream/handle/10568/67909/CCAFSwp.129.pdf?sequence=1

FAO (2011) Land and Water Options for Climate Change Adaptation and Mitigation in Agriculture. Thematic Background Paper for the State of the World's Land and Water Resources for Food and Agriculture (SOLAW)– Managing Systems at Risk. Food andAgriculture Organization of the United Nations.

FAO (2012) Adapting Agriculture to Climate Change, Retrieved from http://www.fao.org/docrep/011/aj982e/aj982e02.pdf

FAO (2013) Climate-Smart Agriculture: Sourcebook. Food and Agriculture Organization of the United nations. Retrieved from http://www.fao.org/3/a-i3325e.pdf

FAO (2016) The State of Food and Agriculture. Food and Agriculture Organization of the United Nations,. Available online: http://www.fao.org/3/a-i6030e.pdf (accessed on 10 April 2019)

Gonsalves J, Sebastian L, Joven B, Amutan C, Lucerna A. 2015. Climate - Smart Villages: Key Concepts. A primer for CCAFS partners in Southeast Asia. International Institute of Rural Reconstruction (IIRR). Copenhagen, Denmark: CGIAR Research Program on Climate Change, Agriculture and Food Security (CCAFS).

https://ccafs.cgiar.org

https://ccafs.cgiar.org/news/how-ccafs-legacy-lives

Jat, M.L., Satyanarayana, T., Majumdar, K., Tetarwal, J.P., Jat, R.K. and Saharawat, Y.S. 2013. 'fertilizer Best Management Practices for Maize systems', Indian Journal of Fertilizer, Vol. 9, April, pp.80–94.

Lal, R. 2004. 'Soil carbon sequestration to mitigate climate change', Geoderma, Vol. 123,Nos. 1–2, pp.1–22, https://doi:org/10.1016/j.geoderma.2004.01.032

NAFCC, 2018. Increasing adaptive capacity to climate change through development of Climate Smart Village (CSV).

Progress for Every Child in the SDG Era» (PDF). UNICEF. Archived (PDF) from the original on 15 July 2020. Retrieved 2 April 2018.

Sapkota, T.B., Majumdar, K., Jat, M.L., Kumar, A., Bishnoi, D.K., McDonald, A.J. and Pampolino, M. 2014. 'Precision nutrient management in conservation agriculture based wheat production of Northwest India : Profitability, nutrient use efficiency and environmental footprint', Field Crops Research, Vol. 155, pp.233–244.

www.google.com

14

Challenges Faced by Farmers in Natural Farming

Soothram Sahithya[1], Prakash Panwar[2]

[1]*Department of Extension Education and Communication Management CCAS, MPUAT, Udaipur, Rajasthan*
[2]*College of Agriculture, Bhilwara, MPUAT, Udaipur, Rajasthan*

Abstract

India has one of the world's largest agricultural production areas of land. The net sown area is 140.1 million hectares, out of which 49 percent is net irrigated area. The proportion of Indians employed in agriculture is decreasing along with its allied sectors such as livestock, forestry and fisheries, still it remains as India's primary source of income. In the 1960s, the country adopted the Green Revolution model. It emphasised on developing new seed varieties as well as increasing the use of synthetic fertilisers to meet the food need of population. This approach succeeded in increasing food production but led to exploitation and degradation of Natural resources of land and water. For this Indian Government encouraged the farmers to do Natural farming (NF). Despite the benefits and opportunities, farmers face significant challenges when attempting to transition to Natural farming system. The objective for the present study is to find out the challenges faced by the farmers in Natural farming and suggestions to overcome those challenges. Purposive convenience sampling was adopted and questionnaire was developed which consist of major challenges faced by famers. The responses were collected by phone call and analysed by using appropriate statistical tools. It was found that the major challenges faced were-lack of awareness, difficulty in Marketing and pricing for the produce. The farmers suggested that, Government should implement schemes and policies to encourage Natural farming, appropriate marketing and pricing for the produce and awareness creation programmes. It can be concluded that Extension functionaries can a take lead to promote Natural farming.

Keywords: *Natural farming, challenges in natural farming, challenges for farmers, agriculture, non-chemical farming.*

Introduction

India has one of the world's largest agricultural production areas of land. The net sown area is 140.1 million hectares, the net irrigated area is 49% of this, which is 68.4 million ha as per 2014–15 land-use statistics (Agriculture Research Data Book 2019). While the proportion of Indians employed in agriculture is decreasing, agriculture, along with its allied sectors such as livestock, forestry and fisheries, remains India's primary source of income. As per the Lok Sabha Standing Committee on Agriculture (2019–20), agriculture and allied industries employed 54.6 percent of India's total workforce. Approximately 70% of rural households still rely primarily on agriculture for a living. Agriculture and allied sectors' share of the country's Gross Value Added (the value of goods and services produced in an area, industry, or sector of an economy) has fallen from 18.2 percent in 2014-15 to 16.5 percent in 2019-20. As per current trends, 60% of India's population (more than 10% of the world's population) will face severe food shortages by 2050.

To resolve this, the country implemented the Green Revolution model—an input and chemical-intensive model in the 1960s to boost food production. The emphasis was on developing new seed varieties simultaneously increased the use of synthetic fertilisers and irrigation to increase food production. This strategy increased food production but failed the test of sustainability. Exploitation and degradation of natural resources such as land and water, biodiversity loss and human and animal health.

One such alternative to chemical fertiliser-based agriculture and high input cost agriculture is zero budget natural farming (ZBNF), a sustainable agricultural system. It exemplifies agroecological principles, with an emphasis on "enhanced soil conditions by managing organic matter and soil biological activity; diversification of genetic resources; enhanced biomass recycling; and enhanced biological interactions". The procedure promotes the complete elimination of synthetic chemical inputs (fertilisers and pesticides) and promotes the use of natural mixtures made from cow dung, cow urine, jaggery, pulse flour, etc., mulching practises, and symbiotic intercropping.

The farmers started to adopt natural farming after knowing its advantages but they are also facing some challenges. This study is conducted to find the challenges faced by the farmers in natural farming.

The objectives for the present study are:

1. To find out the challenges faced by the farmers in Natural farming and
2. To know the suggestions to overcome those challenges

Review of Literature

According to Srishailam *et al.* the major challenges faced by natural farming farmers where labor requirement is increased, Pest management is difficult, non-availability of indigenous cow, Promotion of ZBNF on a large scale without scientific validation and weakened agricultural market infrastructure.

Seufert *et al.,* concluded that, overall, organic yields are typically lower than conventional yields. But those yield differences are highly contextual, depending on system and site characteristics, and range from 5% lower organic yields (rain-fed legumes and perennials on weak-acidic to weak-alkaline soils), 13% lower yields (when best organic practices are used), to 34% lower yields (when the conventional and organic systems are most comparable). Under certain conditions organic systems can thus nearly match conventional yields, whereas under others it at present cannot.

Smith *et al.,*found that, 60% of India's population (>10% of people on Earth) will experience severe food deficiencies by 2050. Increased production is urgently needed, but high costs and volatile prices are driving farmers into debt. Zero budget natural farming (ZBNF) is a grassroots movement that aims to improve farm viability by reducing costs.

Materials and Methodology

Research Design

The exploratory research design is adopted for the present study to study in-depth understanding about the challenges faced by the farmers in natural farming. Quantitative research was used rather than qualitative because quantitative data have closed ended questions and they make the statistical analysis easy.

Location of the Study

Respondents for the present study were selected from the Telangana state of India.

Population

The target population for this study were the farmers doing/practicing natural farming since last 2 years and above.

Sample and Sampling Procedure

Phone call interview method was used for the collection of responses from the farmers doing Natural farming and the sample size was 30 for the present study. The list of farmers practicing natural farming was procured from the local NGO personnel and from the list 30 respondents were selected randomly.

Development of Research Tool

Questionnaire was developed by reviewing many articles which are related to the research problem. The developed questionnaire was given to experts for the evaluation and finalised with the incorporated suggestions and corrections received from the experts. Developed questionnaire consists of two main sections:

General Information

General information included the questionslikename, age, gender, years of experience in natural farming, educational qualification, Net sown area(acres) and contact number

Specific Information

Specific information included challenges that farmers face in Natural farming i.e., Difficulty in Marketing and pricing, Convincing the customers, High pricing for the natural produce, Lack of awareness about natural farming among consumers and Extension functionaries role in promoting natural farming.

Data Collection Procedure

Responses were collected through phone call interview after giving brief introduction about the researcher and explaining farmers about the purpose of doing this research. Firstly, general information was collected. Next, the specific information about challenges in natural farming was recorded through phone call which helped to carry out writing work.

Statistical Tools

Raw data were coded and analysed using the following methods.

Frequencies

In statistics, the frequency (or absolute frequency) of an event is the number of times the event occurred in an experiment or study.

Percentage

Percentage was used for standardization of size by calculating the number of individuals in a given category when the total numbers were 30.

Results

It is evident from the Table 1. that, nearly half (43.33%) of the respondents belonged to 36-50 years followed by 50 years and above (36.67%) and 18-35 years (20.00 %). Majority (90.00 %) of the famers were male and only one-tenth of the respondents were females. Regarding educational qualification, nearly half (40.00 %) of the respondents were graduate and above followed by 12th standard or diploma (26.67%), Up to 10th standard (23.33%), No formal schooling (6.67%) Up to 5th standard (3.33%). Forty percentof the respondents had 2-5 years of experience in natural farming followed by 33.33 percent of respondents had 10 and aboveyears of experience, 6-10 years (26.67%). Nearly half (46.67%) of the farmers had 1-10 acres of net sown area followed by 11-20 acres(36.67%) and 21 and above acres (16.66%).

Table 1: Distribution of respondents according to General Information

n=30

S. No	General information	F	%
1	**Age**		
a.	18-35years	06	20.00
b.	36-50 years	13	43.33
c.	50 years and above	11	36.67
2	**Gender**		
a.	Male	27	90.00
b.	Female	3	10.00
3.	**Educational qualification**		
a.	No formal schooling	02	6.67
b.	Up to 5th standard	01	3.33
c.	Up to 10th standard	07	23.33
d.	12th standard or diploma	08	26.67
e.	Graduate and above	12	40.00
4.	**Years of Experience in Natural Farming**		
a.	2-5 years	12	40.00
b.	6-10 years	08	26.67
c.	10 years and above	10	33.33
5.	**Net sown area**		
a.	1-10 acres	14	46.67
b.	11-20 acres	11	36.67
c.	21 and above acres	05	16.66

Table 2: Challenges faced by the famers in Natural Farming

n=30

S. No.	Challenges	Yes	
		F	%
	Lack of awareness about natural farming in consumers	29	96.67
	Facing difficulties in Marketing and pricing of natural produce	27	90.00
	Convincing the customers to buy the produce	26	86.67
	No value for natural products	30	100.00
	Low yield	29	96.67
	Less role of Extension functionaries	30	100.00

S. No.	Challenges	Yes	
		F	%
	The labour requirement is increased	30	100.00
	Promotion of NF with no scientific data on crop yield	29	96.67
	Pest management is difficult with NF	30	100.00
	Non-availability of indigenous cows	16	53.33
	Disparity between supply and demand	13	43.33
	Confused certification frame work	02	6.67

Discussions and Suggestions

It is clearly evident from the Table 2. that cent percent respondents were facing labour requirement and difficulties in managing pest in natural farming this may be due the 'no use' of chemical fertilizers in farming which leads to face these challenges. The results were supported by Srishailam *et al.* All the respondents (100.00 percent) also faced that, there is no value for the natural produce which is chemical free and highly nutritious when compared to present farming method. This is mainly because lack of awareness about natural farming (96.67%) among the consumers/customers. Farmers facing problems in marketing and pricing due to lack of awareness (90.00 %) among consumers/customers and also feel difficulty in convincing (86.67 %) them for buying the natural produce. The yield with natural farming is low (96.67 %) in the initial years. The results were in harmony with the study of Seufert *et al.* To manage this challenge Government can implement such plans which will give financial assistance to the farmers and farm families in their initial years of natural farming this will encourage the farmers to start natural farming. Promotion of NF with no scientific data on crop yield (96.67 %) to reduce this challenge, Agricultural Universities and research stations need to take an initiative to prove the importance of natural farming with scientific results which will give a great strength to the farmers in adopting the natural farming and selling the produce. Non-availability of indigenous cows was challenge faced by 53.33 percent of the respondents and they expressed that indigenous cows are very much helpful in natural farming. The results were in harmony with the study of Srishailam *et al.* Cow dung is a type of natural manure that is extremely beneficial to plant growth. A single cow's dung is enough to cover five acres of land. It is inexpensive while also preventing land degradation caused by excessive use of chemical fertilisers and preserving soil fertility. Cow's urine is one of the most effective natural insecticides for killing insects and pests that can harm our crops. It is estimated that a single cow's urine can protect a crop covering 10 acres of land. Cow's urine and cow dung are the important ingredients in the preparation of Jivamrita/Jeevamrutha, Bijamrita, Agriastra, Brahmastra and Neemastra. The non-availability of indigenous cows is may be due to Extinction of indigenous cows. Disparity between supply and demand (43.33 %), this happens during the initial years of natural farming. This might be due to the lack of knowledge of customers, the farmers might have grown the crop in large area of agriculture land but there is no demand for the produce it leads disparity between supply and demand. Confused certification frame work (6.67%), this might be due to the farmers selling the produce mostly on the trust/loyalty basis, no customer is asking for the documented certificate proof while purchasing. The awareness about natural farming can be increased with Extension Functionaries Role (100.00%). The Government can take a step towards the initiation of awareness creation programs both rural and urban areas with the help of Extension Functionaries. Government should introduce "Common Cattle Farm" in every village same as like Gram Panchayat and Rythu Vedika, where there is availability of cow dung and urine for farming.

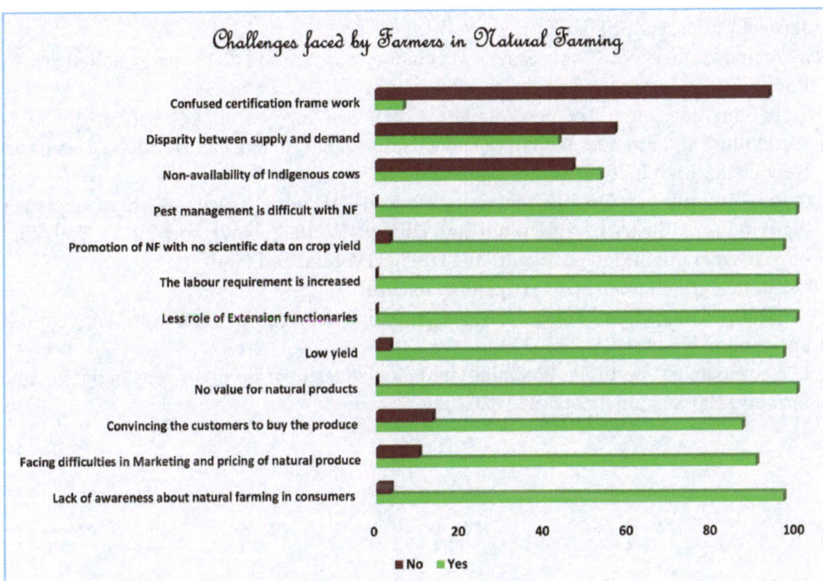

Fig. 1: Challenges faced by Farmers in Natural Farming

Conclusion

With the present study it is concluded that Natural farming is environmentally friendly, low cost, free from chemicals and increases soil fertility. So, if the above discussed major challenges are reduced then there will be a great scope for the high rate of adoption for the natural farming. The challenges can be reduced with policy intervention which are necessary to make success.

Involvement of Government and by conducting awareness creation programs and workshops along with special training programs for the farmers during their initial years of natural farming with help of experts from Agriculture University will lead to an all-round encouragement for youngsters who are interested in Natural Farming.

References

Agriculture Research Data Book 2019, ICAR—Indian Agriculture Statistics Research Institute, Pusa, New Delhi.

Annual report 2018–19, Ministry of Agriculture and Farmers Welfare, Government of India.

Economic Survey 2019–20, Ministry of Finance, Government of India, available at https://www.indiabudget.gov.in/economicsurvey/doc/vol2chapter/echap07_vol2.Pdf

Gupta. N, Tripathi. S, and Dholakia. H. H. 2020. Can Zero Budget Natural Farming Save Input Costs and Fertiliser Subsidies? Evidence from Andhra Pradesh. Council on Energy, Environment and Water. New Delhi:

Khadse, Ashlesha, Rosset,P. M., Morales, H. and Bruce G. Ferguson. 2018. "Taking agroecology to scale: the Zero Budget Natural Farming peasant movement in Karnataka, India." The Journal of Peasant Studies.

Khurana. A and Kumar V., 2020. State of Organic and Natural Farming: Challenges and Possibilities. Centre for Science and Environment. New Delhi.

Ninth report, Standing Committee on Agriculture (2019-20), Seventeenth Lok Sabha, Ministry of Agriculture and Farmers Welfare, Demands for grant (2020–21), Lok Sabha Secretariat, New Delhi, March 2020.

Report of the Committee on Doubling Farmers' Income—Volume V: Sustainability Concerns in Agriculture, Ministry of Agriculture and Farmers Welfare, November 2017, available at http://farmer.gov.in/imagedefault/DFI/DFI%20Volume%205.pdf

Role Of Cow in Agriculture» (tg371) (tractorgyan.com)

Seufert, V., Ramankutty, N. and Foley, J. 2012. Comparing the yields of organic and conventional agriculture. Nature 485, 229–232.

Smith, J., Yeluripati, J., Smith, P. Potential yield challenges to scale-up of zero budget natural farming. Nat Sustain 3, 247–252 (2020).

15

LiFE: Changing Lifestyle for Copping with the Environment and Climate Change

[1]Sushree Purabi Panigrahi, [1]Kalyan Ghadei and [2]Pranoy Roy

[1]Department of Extension Education, Institute of Agricultural Sciences Banaras Hindu University, Varanasi, Uttar Pradesh
[2]SoADU, Bhubaneswar, Odisha

Abstract

Climate change refers to shift in temperature and weather phenomenon. It is supposed to be affected by both natural and manmade factors .The manmade factors contribute more which impacts human health as well as on earth ecosystems. Climate change has influenced our daily life impacting our life style and day to day activities. This review paper highlights how lifestyle contributes to environment and climate change. Our ancient literatures and scriptures highlight how to take care of our mother earth and have co living with nature. Focusing those aspects, our PM's concept of LiFE encompasses on individual behaviourism and it's impact on climate change. This paper is based on the ideas of Narendra Modiji, the prome minister of India of the 17th loksabha (2019-2024).It calls for individuals to make right decisions for our planet and bring behavioural changes even from dinning table. The purpose the paper is to reinforce the idea of LiFE among all citizens of India.

Keywords: LiFE, Behavoural Change, Environment, SDGs, Climate Change

Introduction

The idea of LiFE was introduced by the Prime Minister during the 26th United Nations Climate Change Conference of the Parties (COP26) in Glasgow from 31st October to 13 November 2022. The idea promotes an environmentally conscious lifestyle that focuses on 'mindful and deliberate utilisation' instead of 'mindless and wasteful consumption'.(PIB India)

The LiFE Movement aims to utilise the power of collective action and nudge individuals across the world to undertake simple climate-friendly actions in their daily lives. The LiFE movement, additionally, also seeks to leverage the strength of social networks to influence social norms surrounding climate. The Mission plans to create and nurture a global network of individuals, namely 'Pro-Planet People' (P3), who will have a shared commitment to adopt and promote environmentally friendly lifestyles. Through the P3 community, the Mission seeks to create an ecosystem that will reinforce and enable environmentally friendly behaviours to be self-sustainable.

The Mission envisions replacing the prevalent 'use-and-dispose' economy — governed by mindless and destructive consumption — with a circular economy, which would be defined by mindful and deliberate utilization.

The Prime Minister said that thanks to 1.3 billion Indians in the country he was able to do many good things for the environment in our country. He added that India's forest cover is increasing and so is the population of lions, tigers, leopards, elephants and rhinos. He said that India's commitment to reach 40% of installed electric capacity from non-fossil fuel-based sources has been achieved, 9 years ahead of schedule. The target of 10% ethanol blending in petrol has been achieved 5 months ahead of the November 2022 target. This is a major accomplishment given that blending was hardly 1.5% in 2013-14 and 5% in 2019-20. He said renewable energy has a very high focus in the government. Way ahead, he said, is all about innovation and openness. When technology and tradition mix, the vision of life will be taken further.

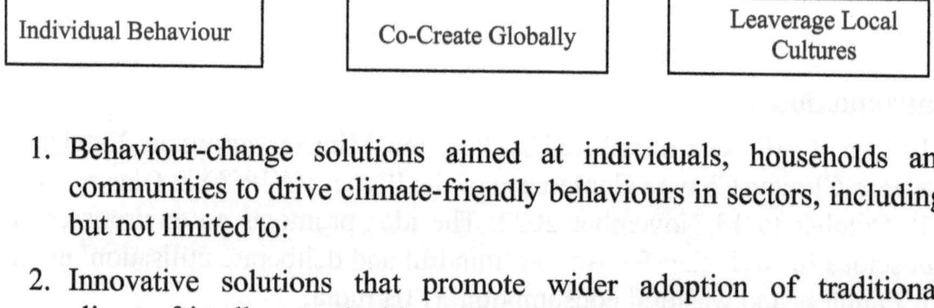

1. Behaviour-change solutions aimed at individuals, households and communities to drive climate-friendly behaviours in sectors, including, but not limited to:
2. Innovative solutions that promote wider adoption of traditional, climate-friendly, sustainable practices and/or create livelihood options for communities that may lose their jobs with a shift towards climate-friendly production.

3. International, national and/or local best practices that can be feasibly scaled-up for driving behaviour change related to climate action.

LiFE is to live a lifestyle that is in tune with our planet and does not harm it. And those who live such a lifestyle are called “Pro-Planet People”. Mission LiFE borrows from the past, operates in the present and focuses on the future. Reduce, Reuse and Recycle are the concepts woven into our life. The Circular Economy has been an integral part of our culture and lifestyle.

The Prime Minister recalled that Mahatma Gandhi talked about a zero-carbon lifestyle. In our daily life choices, let us pick the most sustainable options, he said. He urged the gathering to follow the principle of reuse, reduce and recycle. Our planet is one but our efforts have to be many - One earth, many efforts. “India stands ready to support any effort for a better environment and to further global wellness. Our track record speaks for itself,” .He added that the concept of ‘Reduce, Reuse and Recycle’ and circular economy has been a part of the Indian lifestyle for years. “Such practices are prevalent, which inspire us to walk in harmony with nature. Mission LiFE will encompass every lifestyle related to the conservation of nature, which our ancestors adopted, and that can be made a part of our lifestyle today.”

The programme also witnessed the participation of Mr Bill Gates, Co-Chairman Bill & Melinda Gates Foundation; Lord Nicholas Stern, Climate Economist; Prof. Cass Sunstein, Author of Nudge Theory; Ms. Inger Anderson, UNEP Global Head; Mr Aniruddha Dasgupta, CEO and President World Resources Institute; Mr Achim Steiner, UNDP Global Head; and Mr David Malpass, World Bank President, among others.

“I congratulate PM Modi on leading citizen action and encouraging pro-planet behaviours. Together we can build a green industrial revolution, one that stops climate change, protects vulnerable communities and puts the world on a path to progress. The necessity of collective action to address climate change has never been greater, and India’s role is crucial in ensuring we reach our climate goals,” - Bill Gates, Co-Chairman Bill & Melinda Gates Foundation.

India is the first country to include LiFE in its Nationally Determined Contributions: “India will put forward and propagate a healthy and sustainable way of living based on its traditions and the values of conservation and moderation, including through a mass movement for LIFE, as a key to combating climate change.”Mission LiFE seeks to translate the vision of LiFE into measurable impact. It is designed with the objective to mobilise at least one billion Indians and other global citizens to take individual and collective action for protecting and conserving the environment in the period 2022–28.

Within India, at least 80 percent of all villages and urban local bodies are aimed to become environment-friendly by 2028.

As a global programme, Mission LiFE envisions three core shifts in our collective approach towards sustainability

Change in Demand (Phase I): Nudging individuals across the world to practice simple yet effective environment-friendly actions in their daily lives.

Change in Supply (Phase II): Changes in large-scale individual demand are expected to gradually nudge industries and markets to respond and tailor supply and procurement as per the revised demands.

Change in Policy (Phase III): By influencing the demand and supply dynamics of India and the world, the long-term vision of Mission LiFE is to trigger shifts in large-scale industrial and government policies that can support both sustainable consumption and production.

LiFE and a Sustainable Development Goals (SDGs)

The SDGs focused on sustainable cities and communities (SDG 11), responsible production and consumption (SDG 12), climate change (SDG 13), life on land (SDG 15), and life under water (SDG 14) emphasize that all individuals ensure that their lifestyles are in sync with the resources available on the planet. Further, research from the New Climate Economy highlights that bold environmental action could create as many as 65 million jobs by 2030 (SDG 8: Decent Work and Economic Growth).

Conclusion

The aspiration of living long life free from sickness is depended on our behaviour towards our mother nature. We should be watchful, careful and sacrifice for her. India updated its Nationally Determined Contributions (NDCs) – plans to limit global warming to 1.5 degrees Celsius – in August last year, promising to reduce emissions intensity of GDP by 45% by 2030, from the 2005 level, and achieve 50% cumulative electric power installed capacity from non-fossil fuel-based energy resources by 2030.

References

Ellen MacArthur Foundation (2016). https://ellenmacarthurfoundation.org/circular-economy-in-india

https://data.worldbank.org/indicator/EN.ATM.CO2E.PC?locations=IN-1W

Intergovernmental Panel on Climate Change (2022). https://www.ipcc.ch/report/sixth-assessment-report-working-group-ii/

PIB (2022). https://pib.gov.in/PressReleaseIframePage.aspx?PRID=1860108

SBM Dashboard (accessed on 10th October 2022). https://sbm.gov.in/sbmreport/home.aspx PMUY website (accessed on 10th October 2022). https://pmuy.gov.in/about.html

Swiss Re (2021). https://www.swissre.com/institute/research/topics-and-risk-dialogues/climate-andnatural-catastrophe-risk/expertise-publication-economics-of-climate-change.html

TERI (2022). https://www.teriin.org/sites/default/files/202203/Water%20Efficient%20 Measures%20 for%20Residential%20Townships-compressed_2.pdf R

United Nations Development Programme (2020). https://www.undp.org/belarus/publications/next-frontier-human-development-andanthropocene

United Nations Environment Programme (2020). https://www.unep.org/emissions-gap-report-2020

World Bank data (accessed on 10th October 2022).

16

Climate Change Workforce Development for Extension

Kalyan Ghadei

Professor of Extension Education, Institute of Agricultural Sciences, BHU, Varanasi, Uttar Pradesh

Abstract

This article is prepared from on personal interviews and discussions held with students, scientists, agriculture professionals, and general public and extension agents in agriculture. It is found that increasing awareness and training opportunities for social, economic, educational, and political institutions will create a workforce skilled to adapt and help mitigate the impacts related to climate change. The resulting changes in attitudes and behavior will be the basis for making curricula more relevant, developing coping strategies, and will lead to more profitable and productive climate-smart agriculture.

Keywords: *climate change, agriculture, education, training, workforce, awareness*

Introduction

The intergovernmental panel on climate change (IPCC) which takes account of on historical, psychological, social , physical and economic and political perspectives of climate change in its report, released in November 2014 said that the Working Group I contribution to the IPCC's Fifth Assessment Report (AR5) considers new evidence of climate change based on many independent scientific analyses from observations of the climate system, paleoclimate archives, theoretical studies of climate processes and simulations using climate models. Climate change is not controversial as naysayers and some media would convey. After decades of careful observation, collection of data, and tracking of changes in the climate system, there exists a solid, scientific consensus that human-caused climate change is a reality (Michael, 2012). In their report, the NCSE 2015 says that there is substantial scientific agreement about the occurrence, causes, and consequences of climate change. Yet due to the inherent complexity of the topic and the social controversies surrounding it, confusion and doubt often persist. Here, we need to develop the workforce, education, and training for climate-smart Agriculture. The following are some interventions.

Creation of Awareness

From the discussion made above and review of research it clear that a thorough understanding on climate change is need of the time. It calls for massive level action plan for awareness of different stake holders in Agriculture such as the students, farmers, agri-prenuers, scientists, extension agents and the general public as a whole. A US study reports that five factors that should account for changes in levels of concern: 1) extreme weather events, 2) public access to accurate scientific information, 3) media coverage, 4) elite cues, and 5) movement/ counter movement advocacy. A time-series analysis indicates that elite cues and structural economic factors have the largest effect on the level of public concern about climate change. While media coverage exerts an important influence, this coverage is itself largely a function of elite cues and economic factors. Weather extremes have no effect on aggregate public opinion. Promulgation of scientific information to the public on climate change has a minimal effect. The implication would seem to be that information-based science advocacy has had only a minor effect on public concern, while political mobilization by elites and advocacy groups is critical in influencing climate change concern (Brulle, Carmichael, and Jenkins, 2012). Media coverage of global warming had an immediate but short-term influence on public concern. With such transitory high levels of media coverage we suggest that for more effective communication of climate change, strategies aimed at maintaining mass-media coverage of global warming are required (Yuki and Midori, 2008).

Incorporating Climate Change Issues in School Education

Since the climate is changing we have to bring the issue to the students and the educational system. A well-planned curriculum from school to university education is a need to educate different stakeholders about the impact, adaptation, and mitigation and to bring a profitable climate resilience crop production. The NCSE 2015 urges "It is important for the science of climate change to be taught, both in formal and informal educational environments, in order for future citizens to be able to make scientifically informed decisions about the consequences of climate change. But educators face challenges in helping their students attain climate literacy". The reality of global climate change lends increasing urgency to the need for effective education on earth system sciences, as well as on the human and behavioral dimensions of climate change, from broad societal action to smart energy choices at the household level. It is noteworthy to view of the US government for climate education that explains "the reality of global climate change lends increasing urgency to the need for effective education on earth system sciences, as well as on the human and behavioral dimensions of climate change, from broad societal action to smart energy choices at the household level (NRCNAS,2011).There are other numerous voices from the organization such as UGCRP, ACS, and UNESCO that support both formal and informal climate change education and climate change literacy(NCSE 2015). The IPCC report (2019) finds that strongly reducing greenhouse gas emissions, protecting and restoring ecosystems, and carefully managing the use of natural resources would make it possible to preserve the ocean and cryosphere as a source of opportunities that support adaptation to future changes, limit risks to livelihoods and offer multiple additional societal benefits. IPCC report on October 2019 also describes fellowships and scholarship for motivating students on climate change education.

Workforce Development

Climate change is multidimensional. It needs a workforce of different categories even in Agriculture. We need experts of different types, including climate specialists in every subject in agriculture to overcome climate change. Understanding and responding to climate change, and make suggestions on how we can use the strengths of the human mind and social dynamics to communicate climate change in its temporal context. Then we propose an integration of these insights for the problem of climate change and discuss mismatches between the human mind, surrounding social dynamics, and climate change (Pahl, Sheppard, Boomsma, and Groves, 2014). The demand for clean energy workforce development programs is becoming increasingly important at both the state and local levels. Appropriately trained workers will be needed to transition to a cleaner economy that addresses climate change and energy challenges while also creating new jobs. Some of these new jobs will require little additional training while others will require training or possibly advanced degrees (EPA held a webcast on this topic in January 2010). Building human capacity for improved assessments requires expansion of skills within the existing public and private sectors and developing a much larger workforce that excels at critical and interdisciplinary thinking. The US global change research program suggests useful capacities include the ability to facilitate and communicate research and practice, manage collaborative processes to allow for imaginative analysis and solutions, develop sustainable technologies to reduce climate risks, and build tools for decision-making in an internationally interdependent world. A deeper understanding of the processes and impacts of climate change, disaster risk reduction, energy policy impacts, ecosystem services and biodiversity, poverty reduction, food security, and sustainable consumption requires new approaches to training and curriculum, as well as research to evaluate the effectiveness of different approaches to research and teaching (NCA 2014). The Paryavaran Mitra (Friends of the Environment) in India is an action-based education program for children with the spirit of increasing the Hand Print focusing on Sustainability and Climate Change that envisions creating Paryavaran Mitra in schools. The objective of the program is to guide, facilitate, the capacity to build children through curriculum-linked activities and co-curricular actions in the club mode. Paryavaran Mitra will be groomed to take positive environmental action to have a bearing at individual, community, national and at the global level. The program is based on leveraging partnerships for Education for Sustainable Development action (CCE 2019).

Communication Strategy

The communication strategy is very crucial in climate change extension and its adaptation. It is to decided that what to communicate, when to communicate and to which targeted audience. It needs joint effort by local people to the governmental initiatives to serve a practical, action-oriented guide to communications support to the society.

The strategy should focus on stake holders to be engaged in the process of education and communication, area of changes in Knowledge, attitude and Behavior, selection of message and appropriate channels of communication, communication responsibilities shared by government actors and preparation of procedure and process for fostering better coordination.

The main goal of the strategy should be to encourage a whole-of-government approach to adapting to climate change, so that the Government can act in concert and speaks with one voice. We have to build broad-based public awareness on the Government policies and frameworks on climate change adaptation, convince the public of the need for investment in climate change adaptation from both public and private sources and Support civil action to educate and mobilize on climate change by providing them with tools, resources and opportunities.

According to the report of Department of sustainable development of united states of America (2018) the focus for communications should be on those that will be instrumental in carrying) the message about climate change impacts to the broader public (i.e. journalists, educators, Celebrities and artists), as well as to those sector specific stakeholders (i.e. farmers and fishers) that need information to build their own resilience to climate impacts.

Media: Mainstream local media, regional and international press, business periodicals, trade publications, and blogging sites

Technical experts: Academics and researchers; opinion leaders on environment and sustainable development

Small and medium-sized enterprises: Those individuals and small businesses whose livelihoods are affected by climate change, and need to take steps to protect themselves, such as fishers, farmers and hotel owners

Sports, entertainment and creative arts: Creators (artists and performers) and celebrities who have shown an interest in environmental and social causes, who could become ambassadors on climate change adaptation

Conclusion

The WICC reports that there is a need to take action to offset some of the negative impacts of climate change on specific resources and build capacity on all kinds of stakeholders engaged in agriculture. There is a need to create a better understanding of climate science, impacts and adaptation strategies along with tools for resource managers and other decision-makers. Communicating and establishing a dialog with the public, decision-makers, community groups, local governments, nonprofits, and others about the impacts of climate change and benefits of adaptation. This will help in filling Gaps and expand knowledge about how natural and human systems will respond to climate change. The integration of climate into ongoing outreach programs of agriculture is needed. Further, there is a requirement to develop standardized climate outreach materials for climate literacy.

References

Alejandro Grajal, Susan R. Goldman, and Tracy Marks. (2012) http://climateinterpreter.org/sites/default/files/resources/ClimateChangeEducationEbook First EditionRevised.pdf retrieved

Brulle, R. J., Carmichael, J., & Jenkins, J. C. (2012).Shifting public opinion on climate change: An empirical assessment of factors influencing concern over climate change in the U.S., 2002-2010. Climatic Change, 114, 169–188. doi:10.1007/s10584-012-0403-y

Deressa, T. T., Hassan, R. M., and Ringler C. (2011). Perception and adaptation to climate change by farmers in the Nile basin of Ethiopia. The Journal of Agricultural Science. doi:10.1017/S0021859610000687

Climate Change Education (2015) retrieved from http://ncse.com/climate

Climate change Education at CCE (2019).https://www.ceeindia.org/cee/climate-change-education.html

Clean Energy Workforce Development (2010). Climate Change Adaptation for State and Local Governments retrieved fromhttp://epa.gov/statelocalclimate/web podcasts/local-webcasts.html#wf1

Cooney, C. M. (2010). The perception factor: climate change gets personal. Environmental Health Perspectives. doi:10.1289/ehp.118-a484

Howden, S. M., Soussana, J.-F., Tubiello, F. N., Chhetri, N., Dunlop, M., and Meinke, H. (2007). Adapting agriculture to climate change. Proceedings of the National Academy of Sciences of the United States of America, 104(50), 19691–19696. doi:10.1073/pnas.0701890104

Introduction and Goals of Climate Change Education (2011).Workshop Summary retrieved fromhttp://www.nap.edu/openbook.php?record_id=13224&page=1

IPCC refinement guidelines of 2006 (2019) https://www.ipcc.ch/report/2019-refinement-to-the-2006-ipcc-guidelines-for-national-greenhouse-gas-inventories/

Jamelske, E., Barrett, J., &Boulter, J. (2013). Comparing climate change awareness, perceptions, and beliefs of college students in the United States and China. Journal of Environmental Studies and Sciences. 3(3), 269-278. doi:10.1007/s13412-013-0144-x

Lambert, J. L., &Bleicher, R. E. (2014). Improving climate change communication starting with environmental educators. Journal of Geoscience Education. 62(3), 388-401. doi:10.5408/13-062.1

Legesse, S. A. (2013). Farmers perceptions and adaptation to climate change in Ethiopia. IEEE Potentials, 32(5), 30-33. doi:10.1109/MPOT.2012.2222948

Leiserowitz, A. A. (2005). American risk perceptions: Is climate change dangerous? Risk Analysis, 25(6), 1433-1442. doi:10.1111/j.1540-6261.2005.00690.x

Liu, Z., Smith Jr, W. J., & Safi, A. S. (2014). Rancher and farmer perceptions of climate change in Nevada, USA. Climatic Change, 122(1), 313-327. doi:10.1007/s10584-013-0979-x

Malone, E. L. (2014). Climate perceptions. Issues in Science and Technology, 30(4), 93-93,95. Retrieved from http://ezproxy.lib.vt.edu:8080/login?url=http://search.proquest.com/docview/1649801870?accountid=14826

McNeal, K. S., Spry, J. M., Mitra, R., & Tipton, J. L. (2014). Measuring student engagement, knowledge, and perceptions of climate change in an introductory environmental geology course. Journal of Geoscience Education. 62(4), 655-667. doi:10.5408/13-111.1

Nigatu, A. S., Kloos, H., Asamoah, B. O., Linköpingsuniversitet, Filosofiskafakulteten, Institutionenförtema, & Temavatteninaturochsamhälle. (2014). Knowledge and perceptions about the health impact of climate change among health sciences students in ethiopia: A cross-sectional study. BMC Public Health, 14(587)

National Climate Assessment. U.S. Global Change Research Program (2014) retrieved fromhttp://nca2014.globalchange.gov/report/response-strategies/research-needs#intro-section-2

No denying climate change. (2014). Waterloo Chronicle (Waterloo, Ontario)

Ogunlade, I., Aderinoye-Abdulwahab, S., and Mensah, A. (2014). Knowledge levels of extension agents and their perceived impact of climate change on extension service provision in Ghana. Ethiopian Journal of Environmental Studies and Management. 7(1), 96-103. doi:10.4314/ejesm.v7i1.12

Pahl, S., Sheppard, S., Boomsma, C., & Groves, C. (2014). Perceptions of time in relation to climate change. Wiley Interdisciplinary Reviews-Climate Change, 5(3), 375-388. doi:10.1002/wcc.272

Rejesus, R. M., Mutuc-Hensley, M., Mitchell, P. D., Coble, K. H., & Knight, T. O. (2013). U.S. agricultural producer perceptions of climate change. Journal of Agricultural and Applied Economics, 45(4), 701-701.

Sellmann, D. andBogner, F. X. (2012). Education in global climate change at a botanical garden: Students' perceptions and inquiry-based learning

St Lucias climate change communication strategy (2018) retrieved from (https://climate adapt. eea.europa.eu/metadata/publications/climate-change-communication-strategy-a-west-sussex-case-study

Wisconsin initiatives on climate change impacts retrieved from http://www.wicci.wisc.edu/climate-change.php

Working Group III Contribution to AR5 (IPPCC) Climate Change 2014: Mitigation of Climate Change., retrieved fromhttp://mitigation2014.org/

Writers, S. (2014,). Farmers and scientists divided over climate change. UPI Space Daily

Yuki S., Midori A.U. (2008). Mass-media coverage, its influence on public awareness of climate-change issues, and implications for Japan's national campaign to reduce greenhouse gas emissions. retrieved from http://www.sciencedirect.com/science/article/pii/S0959378008000964.doi:10.1016/j.gloenvcha.2008.10.005

Index